EUCLIDEAN AND
NON-EUCLIDEAN GEOMETRY
An analytic approach

EUCLIDEAN AND NON-EUCLIDEAN GEOMETRY

An analytic approach

PATRICK J. RYAN

McMaster University

CAMBRIDGE
UNIVERSITY PRESS

Published by the Press Syndicate of the University of Cambridge
The Pitt Building, Trumpington Street, Cambridge CB2 1RP
40 West 20th Street, New York, NY 10011-4211, USA
10 Stamford Road, Oakleigh, Melbourne 3166, Australia

First published 1986
Reprinted with corrections 1987
Reprinted 1987, 1988, 1989, 1991, 1992, 1994, 1995

Printed in the United States of America

Library of Congress Cataloging-in-Publication Data is available.

A catalogue record for this book is available from the British Library.

ISBN 0-521-27635-7 paperback

To my teachers
H. S. M. COXETER
and
K. NOMIZU

Contents

Contents

Contents

Preface

This book provides a rigorous treatment of the fundamentals of plane geometry: Euclidean, spherical, elliptic, and hyperbolic. It is intended primarily for upper-level undergraduate mathematics students, since they will have acquired the ability to formulate mathematical propositions precisely and to construct and understand mathematical arguments.

The formal prerequisites are minimal, and all the necessary background material is included in the appendixes. However, it is difficult to imagine a student reaching the required level of mathematical maturity without a semester of linear algebra and some familiarity with the elementary transcendental functions. A previous course in group theory is not required. Group concepts used in the text can be developed as needed.

The book serves several purposes. The most obvious one is to acquaint the student with certain geometrical facts. These are basically the classical results of plane Euclidean and non-Euclidean geometry, congruence theorems, concurrence theorems, classification of isometries, angle addition, trigonometrical formulas, and the like. As such, it provides an appropriate background for teachers of high school geometry.

A second purpose is to provide concrete and interesting realizations of concepts students have encountered or will encounter in their other mathematics courses. All vector spaces are at most three-dimensional, so students do not get bogged down in summation signs and indices. The fundamental notions of linear dependence, basis, linear transformation, determinant, inverse, eigenvalue, and eigenvector all occur in simple concrete surroundings, as do many of the principal ideas of group theory. Also, students will be in a better position to integrate geometry with topology and analysis after having worked with the projective plane and the metric space axioms.

A third purpose is to provide students not only with facts and an understanding of the structure of the classical geometries but also with an arsenal of computational techniques and a certain attitude toward geometrical investigation. They should not be concerned merely with questions of existence (e.g., whether two figures are congruent) but with questions of construction (finding the isometry relating two figures in terms of the given data). Many of the proofs and exercises take this approach. This point of view makes it clearer whether or not a student's "proof" is valid. In addition, it is more appropriate for applications in areas such as computer graphics and computer vision. Although this book does not treat such applications explicitly, the concepts and techniques used are playing an important role in these fast-growing areas of computer science. (See, for example, Foley and van Dam [14], Chapters 7 and 8, and [36].)

The fourth purpose is to provide a link between classical geometry and modern geometry, with the aim of preparing students for further study and research in group theory, Lie groups, differential geometry, topology, and mathematical physics. From this viewpoint the book is actually a study of the real two-dimensional space forms, the (flat) Euclidean plane, the sphere (constant positive curvature), and the hyperbolic plane (constant negative curvature). The isometry groups studied are Lie groups, and the notion of homogeneous space implicitly underlies much of the discussion. Although differential calculus is not used in the book, all the constructs lend themselves easily to differential-geometric treatment.

Our approach to the hyperbolic plane allows one to "do" analytic geometry there without burdensome calculations, thus removing some of the mystery of hyperbolic geometry. Familiarity with the concepts and computational techniques of hyperbolic geometry is an asset to any student of modern geometry and topology. Such central areas as Thurston's work on 3-manifolds and Penrose's work on relativity require a working knowledge of hyperbolic geometry. (See Thurston [32] and Penrose [27] for an introduction to these topics.)

In this book each of the geometries is developed separately. Each geometry has the notion of point, line, distance, perpendicularity, ray, angle, triangle, reflection, congruence, and so forth. The amount of detail with which each topic is treated varies with the setting and generally decreases as the book progresses and readers get their bearings. Some topics have been treated extensively in one setting and very briefly in another. Other topics are merely introduced in the exercises. I have tried to avoid repetition of similar arguments in different settings while allowing readers to see a good variety of methods and viewpoints. For theorems presented in a particular setting, readers should ask themselves the questions: Does the statement make sense in other settings? If so, is it true? Does the same proof work? What modifications are required? Certain unifying notions become evident (e.g., the three reflections theorem, true in all settings for all types of pencils), whereas other statements (e.g., that the perpendicular bisectors of the three sides of a triangle are concurrent) are not universally true, but an appropriate reformulation may be.

The entire book may be covered in a two-semester course. In a one-semester course I usually cover Chapter 1, part of Chapter 2, the first half of Chapter 4, Chapter 7, and those sections of Chapters 5 and 6 relevant to Chapter 7. There are numerous opportunities for excursions into areas of interest to the instructor (e.g., projective geometry, Galilean geometry, Lorentzian geometry, geometry over the complex numbers or finite fields). Some source materials for such excursions are listed in the references.

The exercises are closely related to the text material. Some of them require specific numeric computations and provide a means of testing the students' understanding of the formulas presented. Others require students to supply proofs that have been omitted in the text. Students should do a sufficient number of these so that they are confident that that they could do the others on an examination if required. The rest of the exercises extend the results of the text in some way. They can be omitted without loss of continuity. However, these are the most enjoyable exercises, and students are encouraged to work on as many of them as time permits.

I would like to thank several generations of students at the University of Notre Dame and Indiana University at South Bend, whose interest in geometry provided the impetus for developing the material and whose reactions have helped to shape the book. Thanks are also due to a long list of typists at these institutions and at McMaster University who have worked on the various preliminary versions of the manuscript as well as the final one. Finally, I am grateful to those colleagues and students who have made helpful comments on various versions of the manuscript or assisted in its preparation in other ways: in particular, Nancy Bridgeman, Michael Brown, Kristine Broadhead, Thomas Cecil, H. S. M. Coxeter, K. Nomizu, and Debra van Rie.

PATRICK J. RYAN

Notation and special symbols

General Remarks: Points are denoted by uppercase Roman letters. Lines are denoted by lowercase script letters and lowercase Greek letters. Angles are denoted by uppercase script letters. Figures are usually denoted by uppercase script letters. Some symbols are used in more than one way.

Specific Symbols:

$\mathbf{R}$	the set of real numbers
$\mathbf{R}^2, \mathbf{R}^3, \mathbf{R}^n$	the vector space of all ordered pairs (respectively triples, n-tuples) of real numbers; also called the Cartesian plane (resp. three-space, n-space)
$\mathbf{E}^2, \mathbf{E}^3, \mathbf{E}^n$	the Euclidean plane (resp. three-space, n-space)
$\mathbf{S}^2$	the two-sphere
$\mathbf{P}^2$	the projective plane
$\mathbf{H}^2$	the hyperbolic plane
$\mathbf{D}^2$	the Klein disk model of $\mathbf{H}^2$
$\mathbf{O}(2), \mathbf{O}(3)$	the orthogonal group of $\mathbf{R}^2$ (resp. $\mathbf{R}^3$)
$\mathbf{SO}(2), \mathbf{SO}(3)$	the special orthogonal group of $\mathbf{R}^2$ (resp. $\mathbf{R}^3$)
$\mathbf{GL}(3), \mathbf{GL}(n)$	the group of all invertible 3×3 (resp. $n \times n$) matrices
$\mathbf{SL}(3), \mathbf{SL}(n)$	the special linear group
$\mathbf{PGL}(2)$	the group of collineations of $\mathbf{P}^2$
$\mathbf{GAL}(2)$	the Galilean group of $\mathbf{R}^2$
$\mathbf{S}_3$	the symmetric group of permutations of three letters
$\mathbf{C}_m$	the cyclic group of order m
$\mathbf{D}_m$	the dihedral group of order $2m$
$\mathbf{AF}(2)$	the group of all affine transformations of $\mathbf{R}^2$
$\mathbf{Sim}(\mathbf{E}^2)$	the group of all similarities of $\mathbf{E}^2$
$\mathcal{T}(\mathbf{E}^2)$	the group of all translations of $\mathbf{E}^2$
$\mathcal{I}(M)$	the group of all isometries of a geometry M
$\mathcal{S}(A)$	the group of all symmetries of a set A
$\mathcal{AS}(A)$	the group of all affine symmetries of a set A
$\mathrm{TRANS}(\ell)$	the group of all translations along the line ℓ

REF($\mathscr{P}$)	the group generated by reflections in all lines of the pencil $\mathscr{P}$
ROT(P)	the group of all rotations leaving P fixed
DIS($\mathscr{P}$)	the group of all parallel displacements determined by the pencil $\mathscr{P}$
$\mathscr{P}$	the set of all points
$\mathscr{P}$	a pencil of lines
$\mathscr{L}$	the set of all lines
$\mathscr{F}$	a figure
$d(P, Q)$	the distance between the points P and Q
$d(P, \ell)$	the distance between P and the closest point of ℓ
$d(\ell, m)$	the smallest number of the form $d(P, Q)$ where $P \in \ell$ and $Q \in m$
$\langle x, y \rangle$	the inner product of two vectors in $\mathbf{R}^n$
$\langle A \rangle$	the group generated by a set of elements A
$\lvert a \rvert$	the absolute value of a number a
$\lvert x \rvert$	the length of the vector x in $\mathbf{R}^n$
$[v]$	the set of all multiples of the vector v
$[a]$	the equivalence class to which a belongs
$[a, b]$	the closed interval of real numbers r satisfying $a \leqslant r \leqslant b$
$(a, b]$	the interval of real numbers r satisfying $a < r \leqslant b$
$[v, w]$	the vector space spanned by vectors v and w
$[G{:}H]$	the index of a subgroup H in a group G
$[P; Q \leftrightarrow R]$	the affine reflection of $\mathbf{E}^2$ that leaves P fixed while interchanging Q and R
$[P; Q \rightarrow R]$	the shear of $\mathbf{E}^2$ that leaves P fixed and sends Q to R
$[C; \ell \rightarrow \ell']$	the perspectivity with center C mapping ℓ to ℓ'
(a, b)	the interval of real numbers r satisfying $a < r < b$
(a, b)	the point of $\mathbf{R}^2$ with coordinates a and b
(PQ)	the permutation that interchanges P and Q
PQ	the segment whose end points are P and Q
(PQR)	the cyclic permutation that sends P to Q, Q to R, and R to P
$T(A)$ or TA	the set of all points of the form TP, where $P \in A$
$\overleftrightarrow{PQ}$	the line containing points P and Q
$\overrightarrow{PQ}$	the ray with origin P that contains Q
π	the natural projection determined by an equivalence relation, $\pi(a) = [a]$
π	the number (approximately equal to 3.14) that is the smallest positive number θ satisfying $\sin \theta = 0$
Π	a plane
$\alpha(t)$	parametric representation of a line
$A + B$	the set of all vectors that can be obtained by adding an

element of A to an element of B. If $A = \{P\}$ is a
singleton, $A + B$ may be written $P + B$

$x \in A$	x is an element of the set A
$\varepsilon_1, \varepsilon_2, \ldots, \varepsilon_n$	standard unit basis vectors for $\mathbf{R}^n$
I	the identity matrix; the identity transformation
J	the complex structure on $\mathbf{R}^2$. In fact, $J = \operatorname{rot}(\pi/2)$
$\hat{\mathscr{F}}$	the rectilinear completion of a figure $\mathscr{F}$
$v^{\perp}$	in $\mathbf{R}^2$, if $v = (v_1, v_2)$, then $v^{\perp} = (-v_2, v_1)$
$A^{\perp}$	the set of all vectors orthogonal to every element of A
$\ell \perp m$	ℓ is perpendicular to m
$\ell \parallel m$	ℓ is parallel to m
$\square$	end of a proof
Ω_ℓ	reflection in the line ℓ
τ_v	translation by the vector v
ref θ	reflection in the line through the origin with direction vector $(\cos \theta, \sin \theta)$
rot θ	the rotation about the origin that takes $(1, 0)$ to $(\cos \theta, \sin \theta)$
H_P	the half-turn about P
$u \times v$	the cross product of vectors u and v in $\mathbf{R}^3$
$\measuredangle PQR$	the angle with vertex Q and sides $\overrightarrow{QP}$ and $\overrightarrow{QR}$
$\triangle PQR$	a triangle with vertices P, Q, and R
det A	the determinant of the matrix A
$\det(u, v, w)$	the determinant of the matrix whose rows are u, v, and w
$Gx = \operatorname{Orbit}(x)$	the orbit of a point x by a group G
$G_x = \operatorname{Stab}(x)$	the stabilizer of a point x in a group G
Σ	summation sign
$\#A$	the number of elements in a set A
$\varnothing$	the empty set
A^t	the transpose of the matrix A
$\cong$	congruence of figures
$\cong$	isomorphism of groups

Theorem Numbers: When theorems are referenced outside of the chapter
in which they occur, their numbers are prefixed with the chapter number.
For example, Theorem 4.21 refers to Theorem 21 of Chapter 4.

Historical introduction

0

In the beginning, geometry was a collection of rules for computing lengths, areas, and volumes. Many were crude approximations arrived at by trial and error. This body of knowledge, developed and used in construction, navigation, and surveying by the Babylonians and Egyptians, was passed on to the Greeks. Blessed with an inclination toward speculative thinking and the leisure to pursue this inclination, the Greeks transformed geometry into a deductive science. About 300 B.C., Euclid of Alexandria organized some of the knowledge of his day in such an effective fashion that all geometers for the next 2000 years used his book, *The Elements*, as their starting point.

First he defined the terms he would use – points, lines, planes, and so on. Then he wrote down five postulates that seemed so clear that one could accept them as true without proof. From this basis he proceeded to derive almost 500 geometrical statements or theorems. The truth of these was in many cases not at all self-evident, but it was guaranteed by the fact that all the theorems had been derived strictly according to the accepted laws of logic from the original (self-evident) assertions.

Although a great breakthrough in their time, the methods of Euclid are imperfect by modern standards. To begin with, he attempted to define everything in terms of a more familiar notion, sometimes creating more confusion than he removed. The following examples provide an illustration:

A *point* is that which has no part. A *line* is breadthless length. A straight line is a line which lies evenly with the points on itself. A *plane angle* is the inclination to one another of two lines which meet. When a straight line set upon a straight line makes adjacent angles equal to one another, each of the equal angles is a *right angle*.

Euclid did not define length, distance, inclination, or "set upon." Once having made his definitions, Euclid never used them. He used instead the "rules of interaction" between the defined objects as set forth in his five postulates and other postulates that he implicitly assumed but did not state. Euclid's five postulates were the following:

I. To draw a straight line from any point to any other point.
II. To produce a finite straight line continuously in a straight line.
III. To describe a circle with any center and distance.
IV. That all right angles are equal to each other.
V. That, if a straight line falling on two straight lines makes the interior angles on the same side less than two right angles, the two straight lines, if produced indefinitely, meet on that side on which the angles are less than two right angles.

Euclid did not feel it necessary to enunciate the following postulate, even though he used it in his very first theorem.

Two circles, the sum of whose radii is greater than the distance between their centers, and the difference of whose radii is less than that distance, must have a point of intersection.

It is natural to ask why Euclid singled out his five postulates for explicit mention. After Euclid, mathematicians attempted to make explicit the assumptions that Euclid had neglected to mention. The fifth postulate attracted much attention. It was cumbersome but intuitively appealing, and people felt that it might be deduced from the other assumptions of Euclid. Many "proofs" of the fifth postulate were proposed, but they usually contained a hidden assumption equivalent to what was to be proved. Three such equivalent conditions were:

i. Two intersecting straight lines cannot be parallel to the same straight line. (Playfair)
ii. Parallel lines remain at a constant distance from each other. (Proclus)
iii. The interior angles of a triangle add up to two right angles. (Legendre)

In 1763 a man named Klügel wrote a dissertation at Göttingen in which he evaluated all significant attempts to prove the parallel postulate in the 2000 years since Euclid had stated it. Of the 28 proofs he examined, not one was found to be satisfactory. Of particular interest was the work of the Jesuit Saccheri (1667–1733). Saccheri assumed the negation of the fifth postulate and deduced the logical consequences, hoping to arrive at a contradiction. He derived many strange-looking results, some of which he claimed were inconsistent with Euclid's other postulates. Actually, he had discovered some fundamental facts about what we now call hyperbolic geometry.

Gauss (1777–1855) was apparently the first mathematician to whom it occurred that this negation might never lead to a contradiction and that geometries differing from that of Euclid might be possible. The thought struck him as being so revolutionary that he would not make it public. In 1829 he wrote that he feared the "screams of the dullards," so entrenched were the ideas of Euclid. Lobachevsky (1793–1856) and Bolyai (1802–1860) independently worked out geometries that seemed consistent and yet negated Euclid's fifth postulate. These works were published in 1829 and 1832, respectively. Experience proved that Gauss had overestimated the dullards. They paid no attention to the new theories.

Almost 40 years later Beltrami (1835–1900) and Klein (1849–1925) produced models within Euclidean geometry of the geometry of Bolyai and Lobachevsky (now called *hyperbolic geometry*). It was thus established that if Euclid's geometry was free of contradiction, then so was hyperbolic geometry. Because hyperbolic geometry satisfied all the assumptions of Euclid except the parallel postulate, it was finally determined that a proof of the postulate was impossible.

With this branching of geometry into Euclidean and non-Euclidean, it became useful to categorize results according to their dependence on the fifth postulate. Any theorem of Euclid that made no use of the parallel postulate was called a theorem of *absolute geometry*. It was equally valid in Euclidean and hyperbolic geometry. By contrast, certain Euclidean theorems that depended only on postulates I, II, and V became known as *affine geometry*. Theorems common to absolute and affine geometry are called theorems of *ordered geometry*.

The study of central projection was forced upon mathematicians by the problems of perspective faced by artists such as Leonardo da Vinci (1452–1519). The image made by a painter on canvas can be regarded as a projection of the original onto the canvas with the center of projection at the eye of the painter. In this process, lengths are necessarily distorted in a way that depends on the relative positions of the various objects depicted. How is it possible that the geometric structure of the original can still usually be recognized on the canvas? It must be because there are geometric properties invariant under central projection. *Projective geometry* is the body of knowledge that developed from these considerations. Many of the basic facts of projective geometry were discovered by the French engineer Poncelet (1788–1867) in 1813 while a prisoner of war, deprived of books, in Russia. Affine and projective geometry are also closely related, because the study of those properties of figures that remain invariant under parallel projection also leads to affine geometry. This aspect of affine geometry was recognized by Euler (1707–1783).

Because progress in geometry had been frequently hampered by lack of computational facility, the invention of *analytic geometry* by Descartes (1596–1650) made simple approaches to more problems possible. For instance, it allowed an easy treatment of the theory of conics, a subject which had previously been very complicated. Since the time of Descartes, analytic methods have continued to be fruitful because they have allowed geometers to make use of new developments in algebra and calculus.

The scope of geometry was greatly enlarged by Riemann (1826–1866). He realized that the geometry of surfaces provided numerous examples of new geometries. Suppose that a curve lying on the surface is called a line if each small segment of it is the shortest curve joining its end points. Then, for instance, if the surface is a sphere, the lines are the great circles. In this geometry, called *double elliptic geometry*, the following theorems are valid:

i.   Every pair of lines has two points of intersection. These points are antipodal; that is, they lie at the opposite ends of the same diameter.
ii. Every pair of nonantipodal points determines exactly one line. An antipodal pair has many lines through them.
iii. The sum of the angles of a triangle is greater than π. It is possible for a triangle to have three right angles.

Riemann and Schläfli (1814–1895) considered higher-dimensional Euclidean and spherical spaces, and in his celebrated inaugural lecture at Göttingen in 1854, Riemann laid the foundations of geometry as a study of general spaces of any dimension, which are now called Riemannian manifolds. These spaces are the principal objects of study in modern *differential geometry*. As the name suggests, the methods used depend on calculus. The geometry of Riemann was used by Einstein (1879–1955) as a basis for his general theory of relativity (1916).

Although Gauss observed the relationship between the angle sum of a triangle and the curvature of the surface on which it occurs, Riemann and those who followed him carried these ideas over to Riemannian manifolds. Thus, curvature is still an important phenomenon in differential geometry, and it indicates how much the geometry of the space being studied differs from being Euclidean.

Although Euclid believed that his geometry contained true facts about the physical world, he realized that he was dealing with an idealization of reality. He did not mean that there was such a thing physically as breadthless length. But he was relying on many of the intuitive properties of real objects. In order to free geometry from reliance on physical concepts for its proofs, Hilbert (1862–1943) rewrote the foundations of geometry in 1899. Hilbert started with undefined objects (e.g., points, lines, planes), undefined relations (e.g., collinearity, congruence, betweenness), and certain axioms expressed in terms of the undefined objects and relations. Anything that could be deduced from this by the usual rules of logic was a geometrical theorem valid in that particular geometry. The choice of axioms was a matter of taste. Of course, some geometries would be interesting and some not, but that is a subjective judgment. The theorems do not depend on the nature of the undefined objects but only on the axioms they satisfy.

Seeing all these geometries around him, Klein, in 1872, proposed to classify them according to the groups of transformations under which their propositions remain true. Since then, group theory has been of increasing importance to geometers. The new geometries of Riemann gave rise to complicated groups of transformations. Soon techniques were developed to study these groups in their own right. Much work on the subject was done by Sophus Lie (1842–1899), and these groups became known as *Lie groups* in his honor.

Lie groups and differential geometry are active areas of current mathematical research.

Three approaches to the study of geometry

1. THE AXIOMATIC APPROACH

Following Hilbert, we start with some undefined objects, relations, and an axiom system. Then we deduce the logical consequences. We shall make some use of this approach. However, we need some motivation in order to know which axioms to choose and how to interpret our results. Without this, the study will not be very interesting.

2. THE ANALYTIC APPROACH

A point is represented by an ordered pair, triple, and so forth, of real numbers (or, more generally, elements of some other algebraic structure). Points are defined to be collinear if they satisfy an equation of a certain type. Then every algebraic equation that one can derive will have some geometrical interpretation. In this approach, linear algebra and matrices are used to facilitate computation.

3. THE EMPIRICAL APPROACH

Our goal is to discover geometrical facts about the world we live in. We use only those facts that we can observe and their logical consequences. Thus, one can conceive of trying to discover whether the parallel postulate is true or false in the world of physical space. Gauss, in fact, tried to do this by locating mirrors on three distant mountain peaks and measuring the sum of the angles of the large triangle formed by light rays sent from one peak to another. His results were inconclusive because the limits of experimental error were larger than the deviation of his measurement from π.

An example from empirical geometry

Our experience of the external world comes to us through our senses, especially vision and touch. As we move around and view objects from various places, the objects usually appear to change shape. An important exception is the straight line whose shape appears unchanged by a change

5

in viewpoint. Three aspects of collinearity present themselves. Three points are collinear if they appear to be "in line;" that is, viewing one from another obscures the third. Secondly, if one begins at point A and moves "straight ahead" towards B, one will traverse all points on the line segment AB. Finally, if we stretch an elastic band from A to a suitable close point B, the elastic will fall along the segment AB.

If our main goal were to describe the properties of physical space, it would be valuable to construct and study many axiomatic geometries to see which ones best fit our observation. In physics one can never be completely sure that a certain theory is right. One can only say that it fits the observations better than any other known theory.

A basic question is this: What can we rightfully deduce about the nature of our space by observation? We can shed light on this problem by proposing a hypothetical universe and studying the system from the outside. For a long time it was believed that the earth was flat. However, if we go to a point P in the ocean, sail straight ahead 500 miles to Q, then turn right and go 500 miles to R, then return straight to P, we will find that the distance from R to P is about 667 miles. Checking our results with Pythagoras' theorem, we see that it does not hold for the right-angled triangle PQR. Thus, we see that it is possible to conduct an experiment to show that the geometry of our earth is not Euclidean.

Suppose now that our earth had been a circular cylinder rather than a sphere. If we had performed the same experiment, we would have found that the distance from R to P was about 707 miles, as predicted by Pythagoras. Our experiment would not prove, of course, that our earth was a plane. However, it would not contradict that hypothesis. A more ambitious experiment would be to try to answer the following question. If you start at a point and go straight ahead, is it possible that after a while you will begin to get closer to your starting point? Can you actually reach your starting point in this way?

Nature of the book

Although we will be dealing with many of the aspects of geometry mentioned in the historical sketch, we will not discuss them in chronological order. We will rely heavily on analytic techniques that, of course, were not available to Euclid. The group concept will frequently be used to make our discussions more transparent. Linear algebra, an indispensible tool for any modern treatment of geometry, will be used on almost every page.

The book begins with a thorough investigation of the Euclidean plane. Here we set the pattern for our study of the non-Euclidean geometries. Points, lines, reflections, and distance are defined. Questions of parallelism, perpendicularity, and symmetry are studied. Isometries (distance-preserving transformations) are classified, and the structure of the isometry group is determined.

Many of the facts derived about the Euclidean plane are already familiar to those who have studied geometry from another approach. However, the same format can be used to investigate the projective and hyperbolic planes. The results are beautiful and, in some cases, surprising.

When we have completed our construction of the three consistent models of plane geometry, we will have some appreciation for the kind of experiments in empirical geometry with which two competing models of the universe could be tested. Although we have limited ourselves to the two-dimensional case by studying planes, it is not too hard to see how higher-dimensional Euclidean, elliptic, and hyperbolic spaces could be studied. Modern cosmology attempts to describe the universe in terms of the geometrical properties of a four-dimensional "spacetime." Although discussion of such models is beyond the scope of this book, we hope that the techniques and thought patterns developed by studying this book will be useful to those who might later wish to work in this area. An interesting nontechnical reference is Rucker [28].

For further reading on the ideas discussed in this introduction, readers are referred to Coxeter [8], Faber [13], Greenberg [16], Meschkowski [23], Tietze [30], and Euclid's *Elements* as presented by Heath [18].

1 Plane Euclidean geometry

The coordinate plane

We start with the familiar plane of analytic geometry. Each ordered pair (p_1, p_2) of real numbers determines exactly one *point P* of the plane. The point determined by $(0, 0)$ is called the *origin*.

The ordered pair (p_1, p_2) is also referred to as the *coordinate vector* of P. Although mathematically equivalent, the words "point" and "vector" have different connotations. A vector is usually thought of as a line segment directed from one point to another. We may think of the vector (p_1, p_2) as the line segment beginning at the origin 0 and ending at P. We shall regard the words "point" and "vector" as interchangeable, using whichever suggests the more appropriate picture. The set of all vectors is denoted by $\mathbf{R}^2$.

The vector space $\mathbf{R}^2$

If $x = (x_1, x_2)$ and $y = (y_1, y_2)$, then we define

$$x + y = (x_1 + y_1, x_2 + y_2).$$

If c is a real number and x is a vector, then we define

$$cx = (cx_1, cx_2).$$

These operations are called vector addition and scalar multiplication, respectively. In particular, if $c = -1$, the vector cx is denoted by $-x$.

The vector $0 = (0, 0)$ is called the zero vector. The operations of vector addition and scalar multiplication enjoy the following familiar algebraic properties:

Theorem 1. *For all vectors x, y, and z, and real numbers c and d,*
   i. $(x + y) + z = x + (y + z).$
  ii. $x + y = y + x.$

iii. $x + 0 = x$.
iv. $x + (-x) = 0$.
v. $1x = x$.
vi. $c(x + y) = cx + cy$.
vii. $(c + d)x = cx + dx$.
viii. $c(dx) = (cd)x$.

The inner-product space $\mathbf{R}^2$

Given two vectors x and y, we define

$$\langle x, y \rangle = x_1 y_1 + x_2 y_2.$$

The number $\langle x, y \rangle$ is called the *inner product* of x and y. It is sometimes also called the *dot product* or *scalar product* of x and y.

The following identities concerning the inner product may be easily checked.

Theorem 2.
i. $\langle x, y + z \rangle = \langle x, y \rangle + \langle x, z \rangle$ *for all* $x, y, z \in \mathbf{R}^2$.
ii. $\langle x, cy \rangle = c\langle x, y \rangle$ *for all* $x, y \in \mathbf{R}^2$ *and all* $c \in \mathbf{R}$.
iii. $\langle x, y \rangle = \langle y, x \rangle$ *for all* $x, y \in \mathbf{R}^2$.
iv.   *If* $\langle x, y \rangle = 0$ *for all* $x \in \mathbf{R}^2$, *then* y *must be the zero vector.*

Remark: Theorem 1 says that $\mathbf{R}^2$ is a vector space. Theorem 2 says that the inner product is bilinear, symmetric, and nondegenerate. See Appendix D for further discussion of these notions.

For any vector $x \in \mathbf{R}^2$ we define the length of x to be

$$|x| = \sqrt{x_1^2 + x_2^2}.$$

Note that

$$|x|^2 = \langle x, x \rangle,$$

so that length and inner product are intimately related.

Theorem 3.   *The length function has the following properties:*
i. $|x| \geqslant 0$ *for all* $x \in \mathbf{R}^2$.
ii.   *If* $|x| = 0$, *then* $x = 0$ *(the zero vector).*
iii. $|cx| = |c||x|$ *for all* $x \in \mathbf{R}^2$ *and all* $c \in \mathbf{R}$.

We now state and prove a less immediate property of the inner-product function and its consequence for length.

Theorem 4 (Cauchy–Schwarz inequality).   *For two vectors x and y in $\mathbf{R}^2$ we have*

$$|\langle x, y \rangle| \leq |x||y|.$$

Equality holds if and only if x and y are proportional.

Proof: We restrict our attention to nonzero vectors x and y, the assertion being obviously true when either x or y is zero.

Consider the real-valued function f defined by

$$f(t) = |x + ty|^2 \quad \text{for} \quad t \in \mathbf{R}.$$

Using the properties stated above, we observe that $f(t)$ is nonnegative for all t and that $f(t)$ assumes the value 0 if and only if x is a multiple of y.

On the other hand, f is a polynomial of degree 2. Specifically,

$$f(t) = |x|^2 + 2t\langle x, y \rangle + t^2|y|^2 \tag{1.1}$$

and as such remains nonnegative only if $\langle x, y \rangle^2 \leq |x|^2|y|^2$; that is, $|\langle x, y \rangle| \leq |x||y|$.

In addition, $f(t)$ assumes the value zero only if $|\langle x, y \rangle| = |x||y|$. Thus, $|\langle x, y \rangle| = |x||y|$ if and only if x and y are proportional.   □

Corollary.   *For $x, y \in \mathbf{R}^2$,*

$$|x + y| \leq |x| + |y|. \tag{1.2}$$

Equality holds if and only if x and y are proportional with a nonnegative proportionality factor.

Proof:

$$\begin{aligned}
|x + y|^2 &= |x|^2 + 2\langle x, y \rangle + |y|^2 \\
&\leq |x|^2 + 2|x||y| + |y|^2 \\
&= (|x| + |y|)^2;
\end{aligned} \tag{1.3}$$

hence, $|x + y| \leq |x| + |y|$.

If equality holds here, then we must have

$$\langle x, y \rangle = |x|\,|y|.$$

From our work on the Cauchy–Schwarz inequality, we see that x and y must be proportional. But $x = cy$ leads to

$$\langle x, y \rangle = c\langle y, y \rangle = c|y|^2$$

and

$$|x||y| = |c||y||y| = |c||y|^2.$$

Thus, c must be equal to $|c|$; hence, $c \geq 0$.   □

The plane has both algebraic and geometric aspects. When we think of the algebraic properties, we are thinking of the vector properties of $\mathbf{R}^2$.

We now turn to the geometric concept of distance. If P and Q are points, we define the distance between P and Q by the equation

$$d(P, Q) = |Q - P|.$$

The symbol $\mathbf{E}^2$ will be used to denote the set $\mathbf{R}^2$ equipped with the distance function d.

The concept of distance is a fundamental one in geometry. We will now derive the most important properties of distance. They are stated in the following theorem.

Theorem 5.   *Let P, Q, and R be points of $\mathbf{E}^2$. Then*
 i. $d(P, Q) \geqslant 0$.
 ii. $d(P, Q) = 0$ *if and only if* $P = Q$.
 iii. $d(P, Q) = d(Q, P)$.
 iv. $d(P, Q) + d(Q, R) \geqslant d(P, R)$ *(the triangle inequality)*.

Proof: Because $d(P, Q) = |Q - P| = |-(Q - P)| = |P - Q|$, the first three properties follow from Theorem 3. The fourth property is equivalent to showing that

$$|Q - P| + |R - Q| \geqslant |(Q - P) + (R - Q)| = |R - P|.$$

This, of course, follows from the corollary to Theorem 4. Furthermore, equality holds if and only if $Q - P = u(R - Q)$ for some nonnegative number u. In the next section we will see that this implies that P, Q, and R are collinear. $\square$

Lines

A line in analytic geometry is characterized by the property that the vectors joining pairs of points are proportional. We define a *direction* to be the set of all vectors proportional to a given nonzero vector.

For a given vector v let

$$[v] = \{tv| \ t \in \mathbf{R}\}.$$

If P is any point and v is a nonzero vector, then

$$\ell = \{X| \ X - P \in [v]\} \qquad (1.4)$$

is called the *line* through P with direction $[v]$. See Figure 1.1. We also write (1.4) in the form

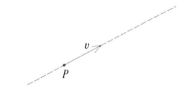

Figure 1.1   The line $\ell = P + [v]$.

$$\ell = P + [v].$$

When $\ell = P + [v]$ is a line, we say that v is a *direction vector* of ℓ.

If ℓ is a line and X is a point, there are many phrases used to express the relationship $X \in \ell$. The following are synonymous:

    i. $X \in \ell$.
    ii. ℓ contains X.
    iii. X lies on ℓ.
    iv. ℓ passes though X.
    v. X and ℓ are incident.
    vi. X is incident with ℓ.
    vii. ℓ is incident with X.

Remark: In axiomatic geometry one usually takes points and lines as fundamental objects and incidence as a fundamental relation. Then an incidence geometry would consist of sets $\mathscr{P}$ and $\mathscr{L}$ and a relation in $\mathscr{P} \times \mathscr{L}$. The relation is assumed to satisfy certain properties from which other properties of the axiomatic system are deduced. See Greenberg [16]. We are being more specific here, but our propositions occur as axioms or propositions in axiomatic developments of plane geometry.

A fundamental property of a line is that it is uniquely determined by any two points that lie on it. Thus, it is important to mention the following:

Theorem 6. *Let P and Q be distinct points of $\mathbf{E}^2$. Then there is a unique line containing P and Q, which we denote by $\overleftrightarrow{PQ}$.*

Proof: Let v be a nonzero vector. The line $P + [v]$ passes through Q if and only if $Q - P \in [v]$. This means that $[Q - P] = [v]$. Hence, the line $P + [Q - P]$ is the unique line required. See Figure 1.2. $\qquad\square$

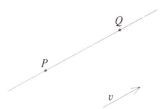

Figure 1.2 The line PQ and a direction vector v.

Thus, a typical point X on the line $\ell = \overleftrightarrow{PQ}$ is written

$$\alpha(t) = P + t(Q - P) = (1 - t)P + tQ. \qquad (1.5)$$

(See Figure 1.3 for a vector addition interpretation.) This equation may be regarded as a parametric representation of the line. As t ranges through the real numbers, $\alpha(t)$ ranges over the line. The parameter is related to distance along ℓ by the formula

$$d(\alpha(t_1), \alpha(t_2)) = |t_2 - t_1|\|Q - P\|. \qquad (1.6)$$

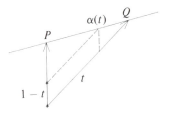

Figure 1.3 $\alpha(t) = (1 - t)P + tQ$.

If $X = (1 - t)P + tQ$, where $0 < t < 1$, we say that X is *between* P and Q. This algebraic characterization of betweenness is equivalent to the following geometrical one.

Theorem 7. *Let P, X, and Q be distinct points of* $\mathbf{E}^2$. *Then X is between P and Q if and only if*

$$d(P, X) + d(X, Q) = d(P, Q).$$

Proof: Suppose first that X is between P and Q. Then for some $t \in (0, 1)$,

$$X = (1 - t)P + tQ.$$

Then

$$d(P, X) = |X - P| = |t(Q - P)| = t|Q - P|.$$

Also,

$$d(X, Q) = |Q - X| = |(1 - t)(Q - P)| = (1 - t)|Q - P|,$$

hence,

$$d(P, X) + d(X, Q) = t|Q - P| + (1 - t)|Q - P|$$
$$= |Q - P| = d(P, Q).$$

Conversely, suppose that X is a point of $\mathbf{E}^2$ satisfying $d(P, X) + d(X, Q) = d(P, Q)$. As we saw in Theorem 5, there is a positive number u such that

$$X - P = u(Q - X).$$

Solving for X gives

$$X = \frac{1}{1 + u}P + \frac{u}{1 + u}Q.$$

Setting $t = u/(1 + u)$, we see that $0 < t < 1$, while $1 - t = 1/(1 + u)$, so that $X = (1 - t)P + tQ$. Thus, X is between P and Q. $\square$

Remark: Theorem 7 is illustrated in Figures 1.4 and 1.5.

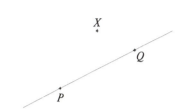

Figure 1.4 $d(P, X) + d(X, Q) > d(P, Q)$. X is not between P and Q.

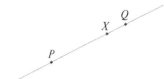

Figure 1.5 $d(P, X) + d(X, Q) = d(P, Q)$. X is between P and Q.

Let P and Q be distinct points. The set consisting of P, Q, and all points between them is called a *segment* and is denoted by PQ. P and Q are the *end points* of the segment. All other points of the segment are called *interior* points.

If M is a point satisfying

$$d(P, M) = d(M, Q) = \tfrac{1}{2}d(P, Q),$$

then M is a *midpoint* of PQ. It follows easily from Exercise 8 that each segment has a unique midpoint, namely,

$$M = \tfrac{1}{2}(P + Q).$$

If two lines ℓ and m pass through a point P, we say that they *intersect* at P and P is their *point of intersection*. From this point of view we restate part of Theorem 6.

13

Theorem 8. *Two distinct lines have at most one point of intersection.*

As we shall see later, two lines in $\mathbf{E}^2$ may have no point of intersection at all.

If three or more lines all pass through a point P, we say that the lines are *concurrent*. If three or more points lie on some line, the points are said to be *collinear*.

Orthonormal pairs

Two vectors v and w are said to be *orthogonal* if $\langle v, w \rangle = 0$. It is frequently desirable to have a vector available that is orthogonal to a given vector. If $v = (v_1, v_2)$, we define $v^\perp = (-v_2, v_1)$. Clearly, v and $v^\perp$ are orthogonal and have the same length. We also easily see that

$$v^{\perp\perp} = -v.$$

A vector of length 1 is said to be a *unit* vector. A pair $\{v, w\}$ of unit orthogonal vectors is called an *orthonormal pair*.

Theorem 9. *Let $\{v, w\}$ be an orthonormal pair of vectors in $\mathbf{R}^2$. Then for all $x \in \mathbf{R}^2$,*

$$x = \langle x, v \rangle v + \langle x, w \rangle w.$$

Proof: Because v and w are linearly independent, they form a basis for $\mathbf{R}^2$ (see Appendix D). Thus, for any $x \in \mathbf{R}^2$, there exist unique constants λ and μ such that $x = \lambda v + \mu w$. But then, using the fundamental properties of the inner product, we get

$$\langle x, v \rangle = \lambda \langle v, v \rangle + \mu \langle w, v \rangle = \lambda$$

and

$$\langle x, w \rangle = \lambda \langle v, w \rangle + \mu \langle w, w \rangle = \mu. \qquad \square$$

Remark: Theorem 9 is illustrated in Figure 1.6.

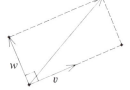

Figure 1.6 Theorem 9.
$x = \langle x, v \rangle v + \langle x, w \rangle w$.

The equation of a line

If ℓ is a line with direction vector v, the vector $v^\perp$ is called a *normal vector* to ℓ. Clearly, any two normal vectors to the same line are proportional. We now derive a characterization of a line in terms of its normal vector. See Figures 1.7 and 1.8.

Theorem 10. *Let P be any point and let {v, N} be an orthonormal pair of vectors. Then $P + [v] = \{X | \langle X - P, N \rangle = 0\}$.*

Proof: By Theorem 9 we have the identity

$$X - P = \langle X - P, v \rangle v + \langle X - P, N \rangle N$$

for any point X in $\mathbf{R}^2$. We show that X lies on the line $P + [v]$ if and only if $\langle X - P, N \rangle = 0$.

First, suppose that $X = P + tv$ for some real number t. Then

$$\langle X - P, N \rangle = \langle tv, N \rangle = t\langle v, N \rangle = 0.$$

Conversely, if $\langle X - P, N \rangle = 0$, the identity reduces to

$$X - P = \langle X - P, v \rangle v,$$

so that

$$X = P + \langle X - P, v \rangle v \ \in P + [v]. \qquad \square$$

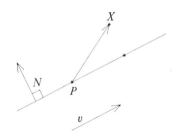

Figure 1.7 $\langle X - P, N \rangle \neq 0$. X does not lie on $\ell = P + [v]$.

Corollary. *If N is any nonzero vector, $\{X | \langle X - P, N \rangle = 0\}$ is the line through P with normal vector N and, hence, direction vector $N^\perp$.*

Proof: Just observe that $\langle X - P, N \rangle = 0$ if and only if $\langle X - P, N/|N| \rangle = 0$ and apply the theorem. $\qquad \square$

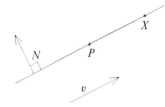

Figure 1.8 $\langle X - P, N \rangle = 0$. X lies on $\ell = P + [v]$.

We recall from elementary analytic geometry that $\{(x, y) | ax + by + c = 0\}$ should represent a line, provided that $a^2 + b^2 \neq 0$. This fits into our scheme as follows:

Theorem 11. *Let a, b, and c be real numbers. Then $\{(x, y) | ax + by + c = 0\}$ is*

i.   *the empty set if $a = 0$, $b = 0$, and $c \neq 0$,*
ii. *the whole plane $\mathbf{R}^2$ if $a = 0$, $b = 0$, and $c = 0$,*
iii. *a line with normal vector (a, b) otherwise.*

Proof: Cases (i) and (ii) are obvious. Consider now the case where $a^2 + b^2 \neq 0$. One can check that the set in question is not empty. In fact, at least one of the points $(-c/a, 0)$ and $(0, -c/b)$ must be defined and satisfy the equation.

Let $P = (x_1, y_1)$ be any point satisfying the equation. Then $c = -(ax_1 + by_1)$. Thus $ax + by + c = 0$ if and only if $a(x - x_1) + b(y - y_1) = 0$. Letting $N = (a, b)$, we see that the set in question is just the line through P with normal vector N. $\qquad \square$

15

Perpendicular lines

Two lines ℓ and m are said to be *perpendicular* if they have orthogonal direction vectors. In this case we write $\ell \perp m$. Two segments are perpendicular if the lines on which they lie are perpendicular.

An important manifestation of perpendicularity is the famous theorem of Pythagoras.

Theorem 12 (Pythagoras). *Let P, Q, and R be three distinct points. Then $|R - P|^2 = |Q - P|^2 + |R - Q|^2$ if and only if the lines $\overleftrightarrow{QP}$ and $\overleftrightarrow{RQ}$ are perpendicular.*

Proof: We recall formula (1.3):

$$|x + y|^2 = |x|^2 + 2\langle x, y \rangle + |y|^2.$$

We note that $|x + y|^2 = |x|^2 + |y|^2$ if and only if $\langle x, y \rangle = 0$. Now put $x = Q - P$ and $y = R - Q$. We see that $x + y = R - P$, and, hence, $|R - P|^2 = |Q - P|^2 + |R - Q|^2$ if and only if $\langle Q - P, R - Q \rangle = 0$. This means that the segment PQ and the segment QR are perpendicular. $\qquad\square$

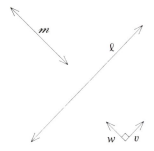

Figure 1.9 Perpendicular lines and their orthonormal direction vectors.

The next property of perpendicular lines is more evident intuitively than Pythagoras' theorem but more difficult to prove. See Figure 1.9.

Theorem 13. *If $\ell \perp m$, then ℓ and m have a unique point in common.*

Proof: Let $\ell = P + [v]$ and $m = Q + [w]$. We may assume that v and w are unit vectors, so that $\{v, w\}$ is an orthonormal set. We write

$$P - Q = \langle P - Q, v \rangle v + \langle P - Q, w \rangle w,$$

and, hence,

$$P - \langle P - Q, v \rangle v = Q + \langle P - Q, w \rangle w.$$

Setting

$$F = P - \langle P - Q, v \rangle v = Q + \langle P - Q, w \rangle w,$$

we see that F lies on both ℓ and m.

F is the only common point, because if there were two, by Theorem 8 the lines would have to coincide. $\qquad\square$

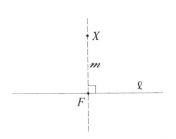

Figure 1.10 Dropping a perpendicular to ℓ from X.

This result allows us to obtain a result motivated by a construction of Euclid.

Theorem 14. *Let X be a point, and let ℓ be a line. Then there is a unique line m through X perpendicular to ℓ. Furthermore,*

i. $m = X + [N]$, *where N is a unit normal vector to ℓ;*
ii. *ℓ and m intersect in the point $F = X - \langle X - P, N\rangle N$, where P is any point on ℓ;*
iii. $d(X, F) = |\langle X - P, N\rangle|.$

Remark: The construction of m when ℓ and X are given is called *erecting a perpendicular to ℓ at X* if X happens to lie on ℓ. Otherwise, it is called *dropping a perpendicular to ℓ from X*. In this case the unique point of intersection of ℓ and m is called the *foot F of the perpendicular*. Theorem 14 is illustrated in Figures 1.10 and 1.11.

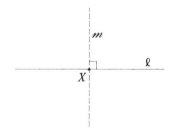

Theorem 15.   *Let ℓ be any line, and let X be a point not on ℓ. Let F be the foot of the perpendicular from X to ℓ. Then F is the point of ℓ nearest to X. (See Figure 1.12.)*

Figure 1.11   Erecting a perpendicular to ℓ at X.

Proof: Let P be any point on ℓ. Because $PF \perp FX$, Pythagoras' theorem gives $|X - P|^2 = |X - F|^2 + |F - P|^2$. Thus, $|X - P|^2 \geq |X - F|^2$ with equality if and only if $P = F$.   □

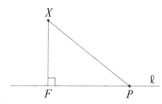

Definition.   *The number $d(X, F)$ is called the* distance from the point X to the line ℓ *and is written $d(X, \ell)$.*

Figure 1.12   *F* is the point of ℓ closest to *X.*

Remark: $d(X, \ell)$ is the shortest distance from X to any point of ℓ.

Corollary.   *Let ℓ be a line with unit normal vector N. Let X be any point of $\mathbf{R}^2$. If P is any point on ℓ, then*

$$d(X, \ell) = |\langle X - P, N\rangle|.$$

We now present another useful construction involving perpendicularity. Let PQ be a segment. The line through the midpoint M of PQ that is perpendicular to $\overleftrightarrow{PQ}$ is called the *perpendicular bisector* of the segment PQ. See Figure 1.13.

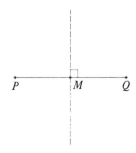

Remark: The perpendicular bisector consists precisely of all points that are equidistant from P and Q.

Figure 1.13   The midpoint and perpendicular bisector of a segment.

Parallel and intersecting lines

Two distinct lines ℓ and m are said to be *parallel* if they have no point of intersection. In this case we write $\ell \parallel m$.

In light of the exercises for the section on lines, if ℓ is any line, P is any point on ℓ, and v is any direction vector of ℓ, then $\ell = P + [v]$.

We have the following criterion for parallelism.

Figure 1.14 $\ell \parallel m$, $m \parallel n$, and $\ell \parallel n$.

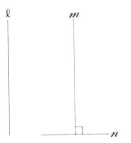

Figure 1.15 $\ell \parallel m$ and $m \perp n$ imply $\ell \perp n$.

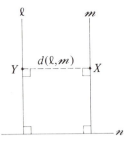

Figure 1.16 $\ell \perp n$, $m \perp n$, and $\ell \parallel m$. $d(X, \ell) = d(Y, m) = d(\ell, m)$.

Theorem 16. *Two distinct lines ℓ and m are parallel if and only if they have the same direction. (Recall that the direction of a line $P + [v]$ is the set $[v]$.)*

Proof: Suppose that ℓ and m have a common point F. We may write $\ell = F + [v]$ and $m = F + [w]$ for nonzero vectors v and w. Because ℓ and m are distinct, $[v] \neq [w]$.

Conversely, suppose that ℓ and m have different directions $[v]$ and $[w]$. Let P be any point of ℓ, and let Q be any point of m. Because v and w are not proportional, there exist numbers t and s such that $P - Q = tv + sw$. (See Appendix D.) This means that $P - tv = Q + sw$. Let $F = P - tv = Q + sw$. Then F is a common point of ℓ and m. $\qquad\square$

Parallel lines come in families, one for each direction. A line m perpendicular to one member ℓ of the family is also perpendicular to all the others. Thus, it is possible to parametrize the family by the real numbers, essentially by measuring distance along m. Although these facts are intuitive and familiar, it is necessary to point them out explicitly here in order to compare them to the analogous situations in non-Euclidean geometries.

We leave the proofs of these to the exercises. Cases (i)–(iii) are illustrated in Figures 1.14–1.16, respectively. Figure 1.16 also illustrates Theorem 18.

Theorem 17.

  i.   If $\ell \parallel m$ and $m \parallel n$, then either $\ell = n$ or $\ell \parallel n$.
  ii. If $\ell \parallel m$ and $m \perp n$, then $\ell \perp n$.
  iii. If $\ell \perp n$ and $m \perp n$, then $\ell \parallel m$ or $\ell = m$.

Theorem 18. *Let ℓ and m be parallel lines. Then there is a unique number $d(\ell, m)$ such that*

$$d(X, \ell) = d(Y, m) = d(\ell, m)$$

for all $X \in m$ and all $Y \in \ell$. In fact, if N is a unit normal vector to ℓ and m, then for any points X on m and Y on ℓ,

$$|\langle X - Y, N \rangle| = d(\ell, m).$$

Thus, parallel lines remain "equidistant." Intersecting lines, on the other hand, behave as follows:

Theorem 19. *Let ℓ be any line, and let m be a line intersecting ℓ at a point P. Let v and w be unit direction vectors of ℓ and m, respectively. Let $\alpha(t) = P + tw$ be a parametrization of m. Then $d(\alpha(t), \ell) = |t| |\langle w, v^\perp \rangle|$. Thus as X ranges through m, $d(X, \ell)$ ranges through all nonnegative real numbers, each positive real number occurring twice. See Figure 1.17.*

Reflections

Any subset of the plane is called a *figure*. Naturally some figures are more interesting than others. Figures with a high degree of symmetry are most interesting, not only because of aesthetic considerations but also because they occur in nature. Snowflakes, molecules, and crystals are three examples of objects with symmetric cross sections.

The simplest kind of symmetry that a plane figure can have is symmetry about a line. See Figure 1.18. We now formulate this notion precisely. Let ℓ be a line passing through a point P and having unit normal N. Two points X and X' are symmetrical about ℓ if the midpoint of the segment XX' is the foot F of the perpendicular from X to ℓ. See Figure 1.19. In other words, X and X' are symmetrical about ℓ if we have

$$\tfrac{1}{2}(X + X') = F.$$

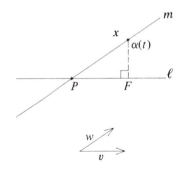

Figure 1.17 $d(a(t), F) = |t| |\langle w, v^{\perp} \rangle|.$

By Theorem 14 this means that

$$\tfrac{1}{2}X + \tfrac{1}{2}X' = X - \langle X - P, N \rangle N,$$

$$\tfrac{1}{2}X' = \tfrac{1}{2}X - \langle X - P, N \rangle N,$$

$$X' = X - 2\langle X - P, N \rangle N.$$

In order to consider symmetry about various lines, it is convenient to adopt a dynamic approach by expressing the relationship between X and X' in terms of a transformation that takes X to X'.

Definition.   *For a line ℓ the reflection in ℓ is the mapping Ω_ℓ of $\mathbf{E}^2$ to $\mathbf{E}^2$ defined by*

$$\Omega_\ell X = X - 2\langle X - P, N \rangle N,$$

where N is a unit normal to ℓ and P is any point of ℓ.

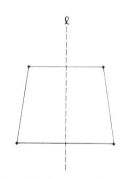

Figure 1.18   A figure that is symmetric about the line ℓ.

If $\mathscr{F}$ is a figure such that $\Omega_\ell \mathscr{F} = \mathscr{F}$, then we say that $\mathscr{F}$ is symmetric about ℓ. The line ℓ is called a line of symmetry or *axis of symmetry* of $\mathscr{F}$.

We now investigate some of the properties of reflections.

Theorem 20.
  i. $d(\Omega_\ell X, \Omega_\ell Y) = d(X, Y)$ *for all points X, Y in $\mathbf{E}^2$.*
 ii. $\Omega_\ell \Omega_\ell X = X$ *for all points X in $\mathbf{E}^2$.*
iii. $\Omega_\ell \colon \mathbf{E}^2 \to \mathbf{E}^2$ *is a bijection.*

Proof:
  i. $\Omega_\ell X - \Omega_\ell Y = X - Y - 2\langle X - Y, N \rangle N.$ Thus,

$$|\Omega_\ell X - \Omega_\ell Y|^2 = |X-Y|^2 - 4(\langle X-Y, N \rangle)^2 + 4(\langle X-Y, N \rangle)^2 \langle N, N \rangle$$
$$= |X - Y|^2.$$

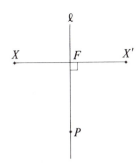

Figure 1.19 X and X' are related by reflection in the line ℓ.

ii. Write $\Omega_\ell X = X - 2\lambda N$, where $\lambda = \langle X - P, N \rangle$. Then

$$\begin{aligned}
\Omega_\ell \Omega_\ell X &= X - 2\lambda N - 2\langle X - 2\lambda N - P, N \rangle N \\
&= X - 2\lambda N - 2\langle X - P, N \rangle N + 4\lambda \langle N, N \rangle N \\
&= X - 2\lambda N - 2\lambda N + 4\lambda N \\
&= X.
\end{aligned}$$

iii. We first show that Ω_ℓ is injective. If $\Omega_\ell X = \Omega_\ell Y$, then $\Omega_\ell \Omega_\ell X = \Omega_\ell \Omega_\ell Y$ and $X = Y$, by (ii). To show that Ω_ℓ is surjective, let Y be any point of $\mathbf{E}^2$. Let $X = \Omega_\ell Y$. Then $\Omega_\ell X = Y$, so that Y is in the range of Ω_ℓ. □

Theorem 21. *$\Omega_\ell X = X$ if and only if $X \in \ell$.*

Proof: $X - 2\langle X - P, N \rangle N = X$ if and only if $\langle X - P, N \rangle N = 0$; that is, $\langle X - P, N \rangle = 0$. The statement now follows from Theorem 10. □

Remark: A *fixed point* of a mapping T is a point X satisfying $TX = X$. Thus, Theorem 21 says that the fixed points of a reflection are those which lie on its axis.

Remark: Theorem 20 shows that reflections are involutive distance-preserving bijections. We study distance-preserving bijections (isometries) in the next section. To say that a mapping T is involutive means $T^2 = TT = I$, the identity mapping of $\mathbf{E}^2$. (See also Appendix B.)

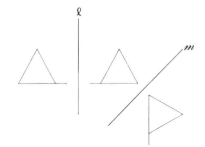

Figure 1.20 Successive reflections $\Omega_m \Omega_\ell$.

Congruence and isometries

If $\mathcal{F}$ is any figure and Ω_ℓ is any reflection, then $\Omega_\ell \mathcal{F}$ is called the *mirror image* of $\mathcal{F}$ in the line ℓ. The figure and its mirror image are observed to have the same "size" and "shape." If Ω_m is a second reflection, then $\Omega_m \Omega_\ell \mathcal{F}$ is again the same size and shape as $\mathcal{F}$. (See Figure 1.20.) One may think of moving $\mathcal{F}$ "rigidly" in the plane so that it coincides with $\Omega_m \Omega_\ell \mathcal{F}$. The key property that makes precise our intuitive notions of size, shape, and rigid motion is that the distance between each pair of points on $\mathcal{F}$ is equal to the distance between the corresponding pairs of points on $\Omega_m \Omega_\ell \mathcal{F}$. We introduce the general concept of distance-preserving mapping or isometry as follows:

Definition. *A mapping T of $\mathbf{E}^2$ onto $\mathbf{E}^2$ is said to be an* isometry *if for any X and Y in $\mathbf{E}^2$,*

$$d(TX, TY) = d(X, Y).$$

Definition. *Two figures $\mathscr{F}_1$ and $\mathscr{F}_2$ are* congruent *if there exists an isometry T such that $T\mathscr{F}_1 = \mathscr{F}_2$.*

We showed in the previous section that every reflection is an isometry. Although not every isometry is a reflection, we shall see later that every isometry is the product (composition) of at most three reflections. Thus, reflections are the basic building blocks of isometries.

Every isometry T is a bijection of $\mathbf{E}^2$ onto $\mathbf{E}^2$. In fact, if $TX = TY$, then

$$0 = d(TX, TY) = d(X, Y),$$

so that $X = Y$. Therefore, the inverse mapping T^{-1} exists. In fact, T^{-1} is also an isometry because

$$d(T^{-1}X, T^{-1}Y) = d(TT^{-1}X, TT^{-1}Y) = d(X, Y).$$

Furthermore, if T and S are isometries, then

$$d(TSX, TSY) = d(SX, SY) = d(X, Y).$$

We now state these results formally.

Theorem 22.
 i. *If T and S are isometries, so is TS.*
 ii. *If T is an isometry, so is T^{-1}.*
 iii. *The identity map I of $\mathbf{E}^2$ is an isometry.*

In other words, the set of all isometries is a group called the *isometry group* of $\mathbf{E}^2$. It is denoted by $\mathscr{I}(\mathbf{E}^2)$.

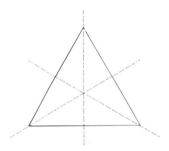

Figure 1.21 An equilateral triangle and its axes of symmetry.

Symmetry groups

Let $\mathscr{F}$ be a figure in $\mathbf{E}^2$. Then the set

$$\mathscr{S}(\mathscr{F}) = \{T \in \mathscr{I}(\mathbf{E}^2) \,|\, T\mathscr{F} = \mathscr{F}\}$$

is a subgroup of $\mathscr{I}(\mathbf{E}^2)$ called the *symmetry group of $\mathscr{F}$*. The fact that $\mathscr{S}(\mathscr{F})$ is a subgroup can be easily verified. The size of the symmetry group of $\mathscr{F}$ is a measure of the degree of symmetry of the figure. We shall show, for example, in a later chapter that an equilateral triangle (Figure 1.21) has a symmetry group of order 6 generated by reflections in the three medians. The isosceles triangle ABC (Figure 1.22) has a symmetry group of order 2 generated by reflection in the median AM. The circle has an infinite symmetry group generated by reflections in all diameters. For an elementary discussion of symmetry groups, see Alperin [2].

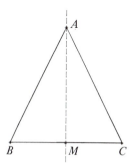

Figure 1.22 An isosceles triangle and its axis of symmetry.

21

Translations

Is there a simple way to describe the product of two reflections? In this section we answer that question affirmatively in the case where the axes of reflection are parallel.

Specifically, if m and n are parallel lines, we may choose P arbitrarily on m and choose Q to be the foot of the perpendicular from P to n. Then if N is a unit normal vector to m (and hence to n), we get

$$
\begin{aligned}
\Omega_m \Omega_n x &= \Omega_n x - 2\langle \Omega_n x - P, N \rangle N \\
&= x - 2\langle x - Q, N \rangle N - 2\langle x - P, N \rangle N + 4\langle x - Q, N \rangle \langle N, N \rangle N \\
&= x + 2\langle P - Q, N \rangle N \\
&= x + 2(P - Q).
\end{aligned}
\tag{1.7}
$$

The last step uses the fact that PQ is perpendicular to m.

Definition. *Let ℓ be any line, and let m and n be perpendicular to ℓ. The transformation $\Omega_m \Omega_n$ is called a* translation *along ℓ. If $m \neq n$, the translation is said to be* nontrivial.

Remark: When two lines in $\mathbf{E}^2$ are perpendicular to ℓ, they are, of course, parallel. On the other hand, when two lines are parallel, there is a line (in fact, infinitely many lines) perpendicular to both. In the geometries we will study later in this book, these properties will fail to be true. In the projective plane, for example, two lines can be perpendicular to a third line but still not be parallel. In the hyperbolic plane, on the other hand, there are parallel lines with no common perpendicular. Thus, if our terminology here seems more specific than necessary, it is being set up so that it will be applicable to the other geometries we study as well.

We now see that in the Euclidean plane, a translation does not determine a line uniquely, although it does determine a parallel family. In the exercises you will be asked to prove the following:

Theorem 23. *Let T be a translation along ℓ. If ℓ' is any line parallel to ℓ, then T is also a translation along ℓ'.*

We also observe that each translation along ℓ has the effect of adding a direction vector of ℓ to each vector in the plane.

Theorem 24. *Let T be a nontrivial translation along ℓ. Then ℓ has a direction vector v such that*

$$
Tx = x + v
\tag{1.8}
$$

for all $x \in \mathbf{E}^2$. Conversely, if v is any nonzero vector and ℓ is any line with direction vector v, then the transformation T determined by (1.8) is a translation along ℓ.

Proof: Let N be a unit direction vector for ℓ. Let P be an arbitrary point of $\mathbf{E}^2$. Let α and β be lines perpendicular to ℓ. (See Figure 1.23.) Let a and b be the unique numbers such that $P + aN \in \alpha$ and $P + bN \in \beta$. Our formula (1.7) becomes

$$\Omega_\alpha \Omega_\beta x = x + 2(P + aN - P - bN)$$
$$= x + 2(a - b)N.$$

If $T \neq I$, we must have $a \neq b$, so that $2(a-b)N$ is the required direction vector.

Conversely, suppose that for each real number λ we define a mapping T_λ by

$$T_\lambda x = x + \lambda N. \tag{1.9}$$

If a and b are any two numbers such that $\lambda = 2(a - b)$, we construct $\alpha = P + aN + [N^\perp]$ and $\beta = P + bN + [N^\perp]$ and observe that $T_\lambda = \Omega_\alpha \Omega_\beta$.

$\square$

We now investigate the group of all isometries generated by reflections in lines perpendicular to ℓ. First we must introduce some new terminology.

Definition. *The set of all lines perpendicular to a given line ℓ in $\mathbf{E}^2$ is called a* pencil of parallels. *The line ℓ is a* common perpendicular *for the pencil. See Figure 1.24.*

We note that taking any line *m* in $\mathbf{E}^2$ together with all lines parallel to *m* would be an equivalent construction.

So far we have discovered that the product of two reflections in lines of a pencil of parallels is a translation along the common perpendicular ℓ. We now investigate further the algebraic structure of the set of isometries formed by reflections of such a family.

We begin with the translations. We denote the set of all translations along ℓ by TRANS(ℓ).

Theorem 25. TRANS(ℓ) *is an abelian group isomorphic to the additive group of real numbers.*

Proof: We adopt the notation of Theorem 24. Then

$$T_\lambda \circ T_\mu(x) = T_\lambda(x + \mu N) = x + \mu N + \lambda N$$
$$= x + (\mu + \lambda)N = T_{\mu + \lambda} x.$$

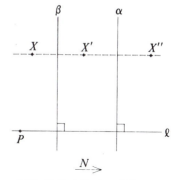

Figure 1.23 $\Omega_\alpha \Omega_\beta$ is the translation along ℓ by an amount equal to twice $d(\alpha, \beta)$. Three successive positions X, X', and X'' of a typical point are shown.

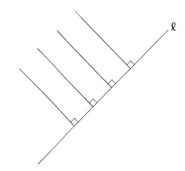

Figure 1.24 A pencil of parallels with common perpendicular ℓ.

Similarly,

$$T_\mu \circ T_\lambda(x) = T_{\lambda+\mu}(x).$$

Because $\lambda + \mu = \mu + \lambda$, translations along ℓ commute.

Further, setting $\lambda = 0$ yields $T_0 = I$ and $T_\lambda \circ T_{-\lambda} = T_0 = I$, so that

$$(T_\lambda)^{-1} = T_{-\lambda}.$$

Thus, TRANS(ℓ) is a subgroup of $\mathscr{I}(\mathbf{E}^2)$. Furthermore, the mapping $\lambda \to T_\lambda$ of $\mathbf{R}$ to TRANS(ℓ) is an isomorphism. This is seen by observing that $T_0 = I$, $T_\lambda^{-1} = T_{-\lambda}$, and $T_\lambda T_\mu = T_{\lambda+\mu}$. $\qquad\square$

Let $\mathscr{P}$ be the pencil of all lines that are perpendicular to a line ℓ. We denote by REF($\mathscr{P}$) the group generated by all reflections of the form Ω_m, where $m \in \mathscr{P}$. In other words, REF($\mathscr{P}$) is the smallest subgroup of $\mathscr{I}(\mathbf{E}^2)$ containing all such Ω_m. In turn, TRANS(ℓ) is a subgroup of REF($\mathscr{P}$).

In order to discuss the algebra of REF($\mathscr{P}$), we need to be able to compute the product of any number of reflections in the family determined by $\mathscr{P}$. We already know that in our notation, $\Omega_\alpha \Omega_\beta = T_\lambda$.

We now take three lines, α, β, γ, of $\mathscr{P}$ corresponding to the numbers a, b, and c. Then

$$
\begin{aligned}
\Omega_\alpha \Omega_\beta \Omega_\gamma &= \Omega_\alpha \circ T_{2(b-c)}, \\
\Omega_\alpha \Omega_\beta \Omega_\gamma x &= \Omega_\alpha(x + 2(b-c)N) \\
&= \Omega_\alpha(x + \mu N), \quad \text{where} \quad \mu = 2(b-c), \\
&= x + \mu N - 2\langle x + \mu N - P - aN, N \rangle N \\
&= x - 2\langle x - P, N \rangle N + (2a - \mu)N \\
&= x - 2\langle x - P, N \rangle N + 2(a - b + c)N \\
&= x - 2\langle x - (P + (a - b + c)N), N \rangle N.
\end{aligned}
$$

We recognize the right side as the formula for reflection in the line $\delta \in \mathscr{P}$ passing through the point $P + dN$, where $d = a - b + c$.

Thus, the product of three reflections in lines of $\mathscr{P}$ is a fourth reflection in a line of the same pencil $\mathscr{P}$. This is our first instance of a *three reflections theorem*, which plays such an important role in classifying the isometries of plane geometries.

Theorem 26 (Three reflections theorem). *Let α, β, and γ be three lines of a pencil $\mathscr{P}$ with common perpendicular ℓ. Then there is a unique fourth line δ of this pencil such that*

$$\Omega_\alpha \Omega_\beta \Omega_\gamma = \Omega_\delta.$$

There are many ways in which a given translation may be represented as the product of two reflections. Using the three reflections theorem, we can exhibit this flexibility precisely.

Theorem 27 (Representation theorem for translations). *Let* $T = \Omega_\alpha \Omega_\beta$ *be any member of* $\mathrm{TRANS}(\ell)$. *If* m *and* n *are arbitrary lines perpendicular to* ℓ, *there exist unique lines* m' *and* n' *such that*

$$T = \Omega_m \Omega_{m'} = \Omega_{n'} \Omega_n.$$

Proof: Apply the three reflections theorem to m, α, and β to produce a unique line m' such that $\Omega_m \Omega_\alpha \Omega_\beta = \Omega_{m'}$. Then multiplying both sides by Ω_m yields $\Omega_\alpha \Omega_\beta = \Omega_m \Omega_{m'}$. The line n' is obtained analogously. ☐

Corollary. *Every element of* $\mathrm{REF}(\mathscr{P})$ *is either a translation along* ℓ *or a reflection in a line of* $\mathscr{P}$.

Proof: This is clear from the following group multiplication table, which summarizes the facts we have established.

	Ω_b	T_μ
Ω_a	$T_{2(a-b)}$	$\Omega_{a-\mu/2}$
T_λ	$\Omega_{b+\lambda/2}$	$T_{\lambda+\mu}$

Here, we have temporarily indexed the reflections Ω by numbers rather than lines. Thus, for example, Ω_a is short for Ω_α, where $\alpha = P + aN + [N^\perp]$. ☐

Let v be any vector in $\mathbf{E}^2$. We define τ_v, the *translation by* v, by

$$\tau_v x = x + v.$$

(If $v = 0$, $\tau_v = I$, and we have the *trivial* translation.) Although all translations arise in the manner we have described, it is possible to discuss in an elementary way the product of two translations that are not along the same line. We get

Theorem 28. *The set* $\mathscr{T}(\mathbf{E}^2)$ *of all translations is an abelian subgroup of* $\mathscr{I}(\mathbf{E}^2)$.

Proof: The following equations are easy to verify and imply the conclusions of the theorem:
1. $\tau_v \tau_w = \tau_{v+w}$.
2. $\tau_0 = I$.
3. $\tau_{-v} = (\tau_v)^{-1}$. ☐

Corollary. $\mathscr{T}(\mathbf{E}^2)$ *is isomorphic to the group* $\mathbf{R}^2$ *with vector addition.*

Rotations

We now investigate the product of reflections in two intersecting lines.

Let $\ell = P + [v]$ be a line with unit direction vector v. There is a unique real number $\theta \in (-\pi, \pi]$ such that

$$v = (\cos\theta, \sin\theta). \quad \text{(See Theorem 1F.)}$$

The unit normal $v^{\perp}$ can be written as

$$N = (-\sin\theta, \cos\theta).$$

We now try to express Ω_ℓ in terms of θ. First note that

$$\Omega_\ell x = x - 2\langle x - P, N \rangle N,$$
$$\Omega_\ell x - P = x - P - 2\langle x - P, N \rangle N.$$

Let ℓ_0 be the line through 0 with direction $[v]$. Then

$$\Omega_{\ell_0} x = x - 2\langle x, N \rangle N.$$

Thus, we see that

$$\Omega_\ell x - P = \Omega_{\ell_0}(x - P),$$

or

$$\Omega_\ell x = \Omega_{\ell_0}(x - P) + P.$$

In other words,

$$\Omega_\ell = \tau_P \Omega_{\ell_0} \tau_{-P}. \quad (1.10)$$

We first deal with Ω_{ℓ_0} and use (1.10) to return to the original situation.

For any x note that

$$\langle x, N \rangle = -x_1 \sin\theta + x_2 \cos\theta.$$

Thus, writing our vectors as column vectors, we get

$$\Omega_{\ell_0}\begin{bmatrix} x_1 \\ x_2 \end{bmatrix} = \begin{bmatrix} x_1 \\ x_2 \end{bmatrix} - 2(-x_1 \sin\theta + x_2 \cos\theta)\begin{bmatrix} -\sin\theta \\ \cos\theta \end{bmatrix}$$

$$= \begin{bmatrix} (1 - 2\sin^2\theta)x_1 + (2\sin\theta\cos\theta)x_2 \\ (2\sin\theta\cos\theta)x_1 + (1 - 2\cos^2\theta)x_2 \end{bmatrix}$$

$$= \begin{bmatrix} \cos 2\theta & \sin 2\theta \\ \sin 2\theta & -\cos 2\theta \end{bmatrix}\begin{bmatrix} x_1 \\ x_2 \end{bmatrix}. \quad (1.11)$$

In other words, $\Omega_{\ell_0}: \mathbf{R}^2 \to \mathbf{R}^2$ is linear. We denote its matrix (see Appendix D) by the symbol ref θ. This matrix represents reflection in the line through the origin whose direction vector is $(\cos\theta, \sin\theta)$:

$$\text{ref } \theta = \begin{bmatrix} \cos 2\theta & \sin 2\theta \\ \sin 2\theta & -\cos 2\theta \end{bmatrix}.$$

We now investigate the matrix algebra of these reflections. First we consider another line m through P and the associated line m_0. Then if $(\cos \phi, \sin \phi)$ is a direction vector of m,

$$\text{ref } \theta \text{ ref } \phi = \begin{bmatrix} \cos 2(\theta - \phi) & -\sin 2(\theta - \phi) \\ \sin 2(\theta - \phi) & \cos 2(\theta - \phi) \end{bmatrix}.$$

We have a special symbol, rot θ, for a matrix of the form

$$\text{rot } \theta = \begin{bmatrix} \cos \theta & -\sin \theta \\ \sin \theta & \cos \theta \end{bmatrix}.$$

Because this linear mapping takes the standard unit basis vector ε_1 to $v = (\cos \theta, \sin \theta)$ and takes ε_2 to $v^\perp = (-\sin \theta, \cos \theta)$, it is reasonable to think of rot θ as a rotation by θ radians in the positive sense. We must keep in mind, however, that definitions of angle, radians, or sense have not yet been given. We now define rotation in such a way that rot θ is a rotation about the origin.

Definition. *If α and β are lines passing through a point P, the isometry $\Omega_\alpha\Omega_\beta$ is called a* rotation about P. *(See Figure 1.25.) The special case $\alpha = \beta$ is allowed so that the identity is (by definition) a rotation about P no matter what P is. If a rotation is not the identity, we refer to it as a* nontrivial *rotation. If $\alpha \perp \beta$, the rotation $\Omega_\alpha\Omega_\beta$ is called a* half-turn.

Theorem 29. *The set of all rotations about the origin is an abelian group called* **SO(2)**.

Proof: Using the formulas from Appendix F, it is easy for us to check that the identities

$$\text{rot } \theta \text{ rot } \phi = \text{rot}(\theta + \phi) = \text{rot}(\phi + \theta) = \text{rot } \phi \text{ rot } \theta,$$

$$\text{rot}(0) = I,$$

$$(\text{rot } \theta)^{-1} = \text{rot}(-\theta)$$

hold. □

The symbol **SO(2)** stands for the special orthogonal group of $\mathbf{E}^2$.

Theorem 30.

  i. $\text{ref } \theta \text{ rot } \phi = \text{ref}\left(\theta - \dfrac{\phi}{2}\right)$.

 ii. $\text{rot } \theta \text{ ref } \phi = \text{ref}\left(\phi + \dfrac{\theta}{2}\right)$.

iii. $\text{ref } \theta \text{ ref } \phi \text{ ref } \psi = \text{ref}(\theta - \phi + \psi)$.

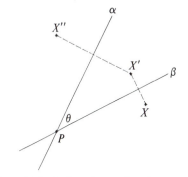

Figure 1.25 $\Omega_\alpha\Omega_\beta$ is the rotation about P by twice θ. Three successive positions X, X', and X'' of a typical point are shown.

Proof: (i)

$$\begin{bmatrix} \cos 2\theta & \sin 2\theta \\ \sin 2\theta & -\cos 2\theta \end{bmatrix} \begin{bmatrix} \cos \phi & -\sin \phi \\ \sin \phi & \cos \phi \end{bmatrix}$$
$$= \begin{bmatrix} \cos(2\theta - \phi) & \sin(2\theta - \phi) \\ \sin(2\theta - \phi) & -\cos(2\theta - \phi) \end{bmatrix}.$$

Equation (ii) is essentially the inverse of (i). Finally, applying (i), we can verify (iii) directly as follows:

$$\text{ref } \theta \text{ ref } \phi \text{ ref } \psi = \text{ref } \theta \text{ rot } 2(\phi - \psi)$$
$$= \text{ref}(\theta - \phi + \psi). \qquad \square$$

Theorem 31. *The set of all rotations about the origin and reflections in lines through the origin is a group called the orthogonal group and is denoted by* **O(2)**. **SO(2)** *is a subgroup of index 2 in* **O(2)**.

Proof: The following group multiplication table is drawn from the facts we have established.

	ref ϕ	rot β
ref θ	rot $2(\theta - \phi)$	$\text{ref}\left(\theta - \dfrac{\beta}{2}\right)$
rot α	$\text{ref}\left(\phi + \dfrac{\alpha}{2}\right)$	$\text{rot}(\alpha + \beta)$

$\square$

The set of reflections is a coset complementary to the coset **SO(2)**.

Let P be any point. The set $\mathscr{P}$ of all lines through P is called the *pencil of lines through P*. We denote by $\text{REF}(\mathscr{P}) = \text{REF}(P)$ the smallest group of isometries containing all Ω_ℓ, where $\ell \in \mathscr{P}$. We denote by $\text{ROT}(P)$ the set of all rotations about P.

Theorem 32. *Let $\mathscr{P}$ be the pencil of all lines through a point P. Then $\text{REF}(\mathscr{P}) \cong$* **O(2)** *and $\text{ROT}(P) \cong$* **SO(2)**.

Proof: For each $T \in$ **O(2)**, $\tau_P \circ T \circ \tau_{-P}$ is in $\text{REF}(P)$. In fact, $\tau_P \text{ ref } \theta \tau_{-P}$ is a reflection, whereas $\tau_P \text{ rot } \theta \, \tau_{-P}$ is a rotation.

It is easy to verify that this provides an isomorphism of **O(2)** onto REF (P) and of **SO(2)** onto $\text{ROT}(P)$. $\square$

We are now ready to prove the analogues of Theorems 26 and 27 for pencils of concurrent lines.

Theorem 33 (Three reflections theorem). *Let α, β, and γ be three lines through a point P. Then there is a unique line δ through P such that*

$$\Omega_\alpha \Omega_\beta \Omega_\gamma = \Omega_\delta.$$

Proof: If

$$\Omega_\alpha = \tau_P(\text{ref } \theta)\tau_{-P}, \quad \Omega_\beta = \tau_P(\text{ref } \phi)\tau_{-P},$$

and

$$\Omega_\gamma = \tau_P(\text{ref } \psi)\tau_{-P},$$

we should choose δ so that

$$\Omega_\delta = \tau_P(\text{ref}(\theta - \phi + \psi))\tau_{-P}.$$

In other words, δ is the line through P with direction vector

$$(\cos(\theta - \phi + \psi), \ \sin(\theta - \phi + \psi)). \qquad \square$$

Theorem 34 (Representation theorem for rotations). *Let $T = \Omega_\alpha \Omega_\beta$ be any member of* ROT(P), *and let ℓ be any line through P. Then there exist unique lines m and m' through P such that*

$$T = \Omega_\ell \Omega_m = \Omega_{m'} \Omega_\ell.$$

Proof: This is similar to the proof for translations. $\qquad \square$

Glide reflections

We now have three basic types of isometries: reflections, translations, and rotations. A fourth type, the *glide reflection*, is defined to be a reflection followed by a translation along the mirror. See Figure 1.26. Specifically, if $\ell = P + [v]$, the glide reflection defined by ℓ and v is given by

$$\tau_v \Omega_\ell x = x - 2\langle x - P, N \rangle N + v,$$

where N is the unit vector $v^\perp/|v|$. Note that

$$\begin{aligned}\Omega_\ell \tau_v x &= x + v - 2\langle x + v - P, N \rangle N \\ &= x + v - 2\langle x - P, N \rangle N\end{aligned}$$

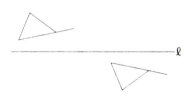

Figure 1.26 A figure and its image by a glide reflection with axis ℓ.

because $\langle v, N \rangle = 0$. Thus, the reflection and translation making up the glide reflection commute. It will be shown that every isometry is one of the four types: reflection, translation, rotation, or glide reflection. Because $\tau_v = I$ is a possibility, each reflection is also a glide reflection. However, glide reflections of this type are said to be *trivial*.

We have dealt with products of reflections in three lines of the same pencil (parallel or concurrent). As an illustration of the power of the tools we have now developed, we analyze the product of reflections in any three lines.

29

Theorem 35. *Let α, β, and γ be three distinct lines that are not concurrent and not all parallel. Then $\Omega_\alpha\Omega_\beta\Omega_\gamma$ is a nontrivial glide reflection.*

Proof: Assume first that α meets β at P. Let ℓ be the line through P perpendicular to γ. Let F be the point of intersection of ℓ and γ. Using the representation theorem for rotations, we know that there is a line m through P such that

$$\Omega_\alpha\Omega_\beta = \Omega_m\Omega_\ell \quad \text{and} \quad \Omega_\alpha\Omega_\beta\Omega_\gamma = \Omega_m\Omega_\ell\Omega_\gamma.$$

Let n be the line through F perpendicular to m, and let n' be the line through F perpendicular to n. Now

$$\Omega_\ell\Omega_\gamma = \Omega_{n'}\Omega_n = H_F,$$

the half-turn about F. As a consequence,

$$\Omega_\alpha\Omega_\beta\Omega_\gamma = \Omega_m\Omega_{n'}\Omega_n.$$

Note that $\Omega_m\Omega_{n'}$ is a translation along n. Because F does not lie on m, n' and m are distinct. Thus, $\Omega_\alpha\Omega_\beta\Omega_\gamma$ is a nontrivial glide reflection.

If α does not meet β but, instead, β meets γ, apply the same argument to $\Omega_\gamma\Omega_\beta\Omega_\alpha = (\Omega_\alpha\Omega_\beta\Omega_\gamma)^{-1}$. If we deduce that $\Omega_\gamma\Omega_\beta\Omega_\alpha = \tau_v\Omega_\ell$, then $\Omega_\alpha\Omega_\beta\Omega_\gamma = (\tau_v\Omega_\ell)^{-1} = \Omega_\ell\tau_{-v} = \tau_{-v}\Omega_\ell$ is also a nontrivial glide reflection. □

Theorem 36. *Let T be a glide reflection, and let Ω_α be any reflection. Then $\Omega_\alpha T$ is a translation or rotation.*

Proof: Let ℓ be the axis of the glide reflection T. There are two cases to consider.

CASE 1: ℓ intersects α. Let P be a point of intersection. By the representation theorem for translations, we may write $T = \Omega_\ell\Omega_a\Omega_b$, where a passes through P, and both a and b are perpendicular to ℓ. Then

$$\Omega_\alpha T = \Omega_\alpha\Omega_\ell\Omega_a\Omega_b.$$

But now α, ℓ, and a all pass through P. By the three reflections theorem there is a line c through P such that

$$\Omega_\alpha T = \Omega_c\Omega_b.$$

Thus $\Omega_\alpha T$ is either a translation or a rotation.

CASE 2: $\ell \parallel \alpha$. Then

$$\Omega_\alpha T = \Omega_\alpha\Omega_\ell\Omega_a\Omega_b = \Omega_\alpha\Omega_a\Omega_\ell\Omega_b.$$

Noting that $b \perp \ell$ and $a \perp \alpha$, we see that $\Omega_\alpha\Omega_a$ and $\Omega_\ell\Omega_b$ are distinct half-turns. By Exercise 26, $\Omega_\alpha T$ is a translation. □

Definition.   *An isometry that is the product of a finite number of reflections is called a* motion.

Theorem 37.   *Every motion is the product of two or three suitably chosen reflections.*

Proof: Suppose that a sequence of reflections is given. If the sequence has length greater than three, we choose any four adjacent elements of the sequence. Applying either Theorems 35 and 36 or one of the three reflections theorems, we can write the product of these four reflections as the product of two. This procedure can be continued until fewer than four reflections remain.   □

Corollary.   *The group of motions consists of all translations, rotations, reflections, and glide reflections.*

Structure of the isometry group

Our main theorem in this section is the following:

Theorem 38.   *Every isometry of* $\mathbf{E}^2$ *is a motion.*

Proof: Let T be an arbitrary isometry. We consider several cases.

CASE 1: $T(0) = 0$. In this case we show in the next lemma that $T = \text{rot } \theta$ or $\text{ref } \theta$ for some value of θ. In other words, $T \in \mathbf{O(2)}$.

CASE 2: $T(P) = P$ for some point P. Then $\tau_{-P}T\tau_P$ is an isometry leaving 0 fixed and is, hence, a member of $\mathbf{O(2)}$ by Case 1. Thus, $T = \tau_P(\text{rot } \theta)\tau_{-P}$ or $T = \tau_P(\text{ref } \theta)\tau_{-P}$. In either case, T is a motion.

CASE 3. T has no fixed points. Let $P = T(0)$. Then $\tau_{-P} \circ T$ leaves 0 fixed and is, hence, either $\text{rot } \theta$ or $\text{ref } \theta$. In any case, $T = \tau_P \text{ rot } \theta$ or $T = \tau_P \text{ ref } \theta$, so that T is a motion.   □

Remark: Case 2 could have been handled as part of Case 3. However, the representation obtained this way is more useful.

We now prove the lemma referred to in Case 1. Although the notion of isometry depends only on distance, the lemma shows that an isometry leaving the origin fixed has a particularly nice algebraic form – it must be linear.

31

Lemma. *If T is an isometry with $T(0) = 0$, then*

i. $\langle Tx, Ty \rangle = \langle x, y \rangle$.

ii. $T = \operatorname{rot} \theta$ *or* $T = \operatorname{ref} \theta$ *for some* θ.

Proof:

i.   By the polarization identity (see Exercise 29),

$$\langle x, y \rangle = \tfrac{1}{2}(|x|^2 + |y|^2 - |x - y|^2),$$

$$\langle Tx, Ty \rangle = \tfrac{1}{2}(|Tx|^2 + |Ty|^2 - |Tx - Ty|^2).$$

Now

$$|Tx| = d(0, Tx) = d(T(0), Tx)$$
$$= d(0, x) = |x|.$$

Similarly,

$$|Ty| = |y|.$$

Also

$$|Tx - Ty| = d(Tx, Ty) = d(x, y) = |x - y|.$$

ii. Let $\varepsilon_1 = (1, 0)$ and $\varepsilon_2 = (0, 1)$. If $x = x_1\varepsilon_1 + x_2\varepsilon_2$, then $\{T\varepsilon_1, T\varepsilon_2\}$ is an orthonormal basis for $\mathbf{E}^2$. Hence,

$$Tx = \langle Tx, T\varepsilon_1 \rangle T\varepsilon_1 + \langle Tx, T\varepsilon_2 \rangle T\varepsilon_2.$$

Using the result of (i), we obtain

$$Tx = \langle x, \varepsilon_1 \rangle T\varepsilon_1 + \langle x, \varepsilon_2 \rangle T\varepsilon_2 = x_1 T\varepsilon_1 + x_2 T\varepsilon_2.$$

Now $T\varepsilon_1$ is a unit vector. Writing

$$T\varepsilon_1 = \lambda_1 \varepsilon_1 + \lambda_2 \varepsilon_2,$$

we see by Theorem 9 and the Cauchy–Schwarz inequality that

$$|\lambda_1| = |\langle T\varepsilon_1, \varepsilon_1 \rangle| \leq |T\varepsilon_1||\varepsilon_1| = 1.$$

Similarly,

$$|\lambda_2| \leq 1 \quad \text{and} \quad |T\varepsilon_1|^2 = \lambda_1^2 + \lambda_2^2,$$

so that

$$\lambda_1^2 + \lambda_2^2 = 1.$$

As in our discussion of rotations, there is a unique $\theta \in (-\pi, \pi]$ such that $\lambda_1 = \cos \theta$ and $\lambda_2 = \sin \theta$. Now

$$\langle T\varepsilon_2, T\varepsilon_1 \rangle = \langle \varepsilon_2, \varepsilon_1 \rangle = 0,$$

so that

$$T\varepsilon_2 = \pm(T\varepsilon_1)^{\perp}.$$

In other words,

$$T\varepsilon_2 = \pm((-\sin\theta)\varepsilon_1 + (\cos\theta)\varepsilon_2).$$

Writing this in matrix form, we have that either

$$Tx = \begin{bmatrix} \cos\theta & -\sin\theta \\ \sin\theta & \cos\theta \end{bmatrix}\begin{bmatrix} x_1 \\ x_2 \end{bmatrix} = (\text{rot }\theta)x$$

or

$$Tx = \begin{bmatrix} \cos\theta & \sin\theta \\ \sin\theta & -\cos\theta \end{bmatrix}\begin{bmatrix} x_1 \\ x_2 \end{bmatrix} = \left(\text{ref }\frac{\theta}{2}\right)x$$

for all $x \in \mathbf{E}^2$. □

Fixed points and fixed lines of isometries

Theorem 39.
  i.   *A nontrivial translation has no fixed points.*
  ii. *A nontrivial rotation has exactly one fixed point, the center of rotation.*
  iii. *A reflection has a line of fixed points, the axis of reflection.*
  iv. *A nontrivial glide reflection has no fixed points.*
  v.   *The identity has a plane of fixed points.*

Proof: Suppose that T is an isometry with no fixed points. As we have shown in the previous section, $T = \tau_P \text{ rot }\theta$ or $T = \tau_P \text{ ref }\theta$ for suitable P and θ. In the first case

$$Tx = (\text{rot }\theta)x + P$$

for all x, so that $Tx = x$ if and only if $(I - \text{rot }\theta)x = P$. But

$$\det(I - \text{rot }\theta) = (1 - \cos\theta)^2 + \sin^2\theta$$

$$= 2 - 2\cos\theta = 4\sin^2\frac{\theta}{2}.$$

Now $\sin^2(\theta/2) = 0$ if and only if rot $\theta = I$. Thus, unless rot $\theta = I$, the equation $(I - \text{rot }\theta)x = P$ has a solution (see Appendix D), and thus T has a fixed point. Because T has no fixed point, rot $\theta = I$ and $T = \tau_P$. Conversely, of course, a nontrivial translation has no fixed point.

We now examine the second case, $T = \tau_P \text{ ref }\theta$. Observe that T is the product of three reflections. Using Theorem 35, we see that T is a nontrivial glide reflection. Conversely, a nontrivial glide reflection can have no fixed points. For if $\Omega_\ell \tau_v$ is a glide reflection with $\ell = P + [v]$, and

$$x = \Omega_\ell(x + v) = x + v - 2\langle x + v - P, N\rangle N,$$

then, because $\langle v, N\rangle = 0$,

$$v = 2\langle x - P, N\rangle N.$$

But then

$$\langle v, v \rangle = 2\langle x - P, N \rangle \langle N, v \rangle = 0,$$

so that $v = 0$.

Thus, an isometry has no fixed points if and only if it is a nontrivial translation or glide reflection. In particular, statements (i) and (iv) hold.

Let T be an isometry with just one fixed point P. From Case 2 of the previous section, T is a rotation about P or a reflection in a line through P. By Theorem 21 the fixed point set of a reflection consists of the axis of reflection itself. Hence, T must be a rotation. Conversely, a nontrivial rotation has exactly one fixed point. For if

$$T = \tau_P(\text{rot } \theta)\tau_{-P},$$

then x is a fixed point if and only if

$$x = P + (\text{rot } \theta)(x - P);$$

that is,

$$(I - \text{rot } \theta)(x - P) = 0.$$

Again, because $\det (I - \text{rot } \theta) \neq 0$, the only possibility is $x - P = 0$; that is, $x = P$. Thus, an isometry has exactly one fixed point if and only if it is a nontrivial rotation. This implies (ii).

Statement (iii) is just Theorem 21, and statement (v) is trivially true. This completes the proof. $\square$

Corollary. *The fixed point set of an isometry must be one of the following:*
  i. *a point (rotation)*
  ii. *a line (reflection)*
  iii. *the empty set (translation or glide reflection)*
  iv. *the whole plane $\mathbf{E}^2$ (the identity).*

If ℓ is any line and T is an isometry, then $T\ell$ is a line, as we will see in Chapter 2. An isometry induces a bijection $T: \mathscr{L} \to \mathscr{L}$ of the set $\mathscr{L}$ of lines onto itself. If ℓ is a line such that $T\ell = \ell$, we say that ℓ is a *fixed line* of T. It is useful to classify the isometries of $\mathbf{E}^2$ with respect to their fixed lines.

Theorem 40.
  i. *A nontrivial translation along a line ℓ has a pencil of parallels as its fixed lines. This pencil consists of all lines parallel to ℓ.*
  ii. *A half-turn centered at C has the pencil of lines through C as its set of fixed lines. A nontrivial rotation that is not a half-turn has no fixed lines.*
  iii. *A reflection Ω_m has the line m and its pencil of common perpendiculars as its fixed lines.*

iv. *A nontrivial glide reflection has exactly one fixed line – its axis.*
v. *The identity leaves all lines fixed.*

When trying to understand the effect of a particular isometry, de-
termination of its fixed points and fixed lines is a good starting point. These
notions and techniques apply in a wider context and will be pursued further
in later chapters.

We will defer the proof of Theorem 40 until we have developed more
convenient algebraic machinery.

EXERCISES

1. Prove Theorem 1.

2. Prove Theorem 2.

3. Prove Theorem 3.

4. Fill in the details required to obtain the expression for $f(t)$ in formula
    (1.1).

5. Show that the result of the corollary to Theorem 4 can be used to
    obtain the inequality

    $$|x| - |y| \leq |x - y|.$$

6. Although P and v determine a unique line ℓ, show that ℓ does not
    determine P or v uniquely.

7. If $\ell = P + [v] = Q + [w]$, how must P, Q, v, and w be related?

8. If $0 < t < 1$ and $X = (1 - t)P + tQ$, and $P \neq Q$, show that

    $$\frac{d(P, X)}{d(X, Q)} = \frac{|P - X|}{|X - Q|} = \frac{t|P - Q|}{(1 - t)|P - Q|} = \frac{t}{1 - t}.$$

    Use this to find the point X that divides the segment PQ in the ratio
    $r{:}s$. Illustrate using $r = 2$, $s = 3$, $P = (-3, 5)$, $Q = (8, 4)$.

9. If v is a nonzero vector, show that there are exactly two unit vectors
    proportional to v.

10. Find an orthonormal pair one of whose members is proportional to
    $(4, -3)$.

11. i. Find all unit normal vectors to the line $3x + 2y + 10 = 0$.
    ii. Find all unit direction vectors of the same line.
    iii. If $P = (5, 2)$ and $v = (\tfrac{1}{2}, \tfrac{2}{3})$, find the equation of the line $P + [v]$
        in the form $ax + by + c = 0$.

12. If $v = (v_1, v_2)$ is a direction vector of a line ℓ, the number $\alpha = v_2/v_1$ is
    called the slope of ℓ, provided that $v_1 \neq 0$.
    i. Show that the concept of slope is well-defined.

35

ii. Show that if ℓ is a line with slope α, the vector $(1, \alpha)$ is a direction vector of ℓ.

iii. Show that the line through $P = (x_1, y_1)$ with slope α has the equation

$$y - y_1 = \alpha(x - x_1).$$

13. $d(X, \ell)$ seems to depend on the choice of P on ℓ and on the unit normal vector N. Show that if N' is another unit normal vector to ℓ and if P' is another point on ℓ, then

$$|\langle X - P, N \rangle| = |\langle X - P', N' \rangle|.$$

14. Let $P + [v]$ and $Q + [w]$ be intersecting lines. Let D be the matrix whose first row is v and whose second row is w. If $P - tv = Q + sw$ is the point of intersection, prove that $(t, s) = (P - Q)D^{-1}$. Here (t, s) and $P - Q$ are regarded as 1×2 matrices. Use this method to find the intersection point in the case $P = (1, 5)$, $Q = (3, 7)$, $v = (8, 1)$, $w = (6, 2)$.

15. Prove Theorems 17–19.

16. Let ℓ and m be parallel lines. Let

$$n = \{\tfrac{1}{2}(X + Y) | X \in \ell \quad \text{and} \quad Y \in m\}.$$

Prove that n is a line parallel to ℓ and m and lying midway between them. In other words, $d(m, n) = d(\ell, n)$.

17. The definition of Ω_ℓ seems to depend on P and N. Show that if P' is another point on ℓ and N' is any unit normal to ℓ, then, for all points X,

$$\langle X - P, N \rangle N = \langle X - P', N' \rangle N'.$$

(Compare with Exercise 13.)

18. Prove Theorem 23.

19. Let $\mathscr{P}$ be a pencil of parallels as discussed in Theorems 25–27.
   i. Show that $\mathrm{REF}(\mathscr{P})$ is isomorphic to the multiplicative group of 2×2 matrices of the form

$$\begin{bmatrix} \pm 1 & \rho \\ 0 & 1 \end{bmatrix},$$

where the reflection Ω_a corresponds to the matrix

$$\begin{bmatrix} -1 & 2a \\ 0 & 1 \end{bmatrix},$$

and T_λ corresponds to

$$\begin{bmatrix} 1 & \lambda \\ 0 & 1 \end{bmatrix}.$$

ii. Observe that TRANS(ℓ) is a subgroup of index 2 in REF($\mathscr{P}$).

20. Verify the statements in Theorem 28 and its corollary.

21. Prove that
i. $\mathscr{T}(\mathbf{E}^2)$ is a normal subgroup of $\mathscr{I}(\mathbf{E}^2)$.
ii. If ℓ is a line, TRANS(ℓ) is not a normal subgroup of $\mathscr{I}(\mathbf{E}^2)$.

22. Let $\ell = P + [v]$ be a line. Let $m = Q + [v]$. Show that if $|v| = 1$, then

$$\Omega_\ell \Omega_m = \tau_w, \quad \text{where} \quad w = 2\langle P - Q, v^\perp \rangle v^\perp,$$

and

$$\Omega_m \Omega_\ell = \tau_{-w}.$$

23. Let τ_w be any translation. Let $\ell = P + [w^\perp]$ be any line having w as a normal vector. Show that if $m = P - \frac{1}{2}w + [w^\perp]$ and $m' = P + \frac{1}{2}w + [w^\perp]$, we have

$$\Omega_\ell \Omega_m = \Omega_{m'} \Omega_\ell = \tau_w.$$

24. Show that the identity is the only rotation that can be described as a rotation about two different points. The unique point P determined by a given nontrivial rotation is called the *center of rotation*.

25. Verify the statements made in the proof of Theorem 32.

26. i. Show that two distinct reflections Ω_ℓ and Ω_m commute if and only if $m \perp \ell$.
ii. Let P be any point. Prove that the half-turn about P is given by

$$H_P x = -x + 2P \quad \text{for all} \quad x \in \mathbf{E}^2.$$

iii. Show that the product of two distinct half-turns is a translation along the line joining their centers.

27. If H_1, H_2, and H_3 are half-turns, prove that

$$H_1 H_2 H_3 = H_3 H_2 H_1.$$

28. Describe the product of two glide reflections whose axes are parallel.

29. The polarization identity

$$\langle x, y \rangle = \tfrac{1}{2}(|x|^2 + |y|^2 - |x - y|^2)$$

allows us to express the inner product in terms of lengths. Prove it.

30. We know that each element of $\mathscr{I}(\mathbf{E}^2)$ can be written uniquely in the form $\tau_P \alpha$, where $\alpha \in \mathbf{O(2)}$ and $P \in \mathbf{R}^2$. Show that the function $\tau_P \alpha \to \alpha$ is a homomorphism of $\mathscr{I}(\mathbf{E}^2)$ onto $\mathbf{O(2)}$. What is the kernel?

31. Prove that the matrix rot θ has a nonzero eigenvalue if and only if rot $\theta = \pm I$.

32. Let α be an isometry such that $\alpha^n = I$. If n is an odd integer, what can you say about α? Explain.

33. Describe the group generated by K in the following cases:
    i. $K = \{\tau_v\}$.
    ii. $K = \Omega_\ell$.
    iii. $K = \{\Omega_\ell, \Omega_m\}, \quad \ell \perp m$.
    iv. $K = \{\tau_v, \tau_w\}$.
    v. $K = \{\tau_v \circ \Omega_\ell\}, \quad \ell = P + [v]$.

34. If ℓ, m, and n are lines of a pencil, prove that $\Omega_\ell \Omega_m \Omega_n = \Omega_n \Omega_m \Omega_\ell$.

35. Let ρ be a nontrivial rotation with center P. Let v be any vector. Show that $\tau_v \rho$ is a rotation. Find its center in terms of the given information.

36. Let P, Q, R, and S be four points, no three of which are collinear. Let A, B, C, and D be the respective midpoints of the segments PQ, QR, RS, and SP. Prove that $\overleftrightarrow{AB} \parallel \overleftrightarrow{CD}$ and $\overleftrightarrow{AD} \parallel \overleftrightarrow{BC}$ or they coincide.

37. i. Prove the remark following the definition of perpendicular bisector.
    ii. Find the perpendicular bisector of the segment joining $(-2, 6)$ and $(4, 8)$.

Affine transformations in the Euclidean plane

2

Affine transformations

In Chapter 1 we discussed the two fundamental aspects of geometry: the incidence aspect based on the notion of collinearity, and the metric aspect based on the notion of distance. Isometries are the transformations that respect these features.

In this chapter we want to enlarge our world of transformations to include those that respect incidence but do not necessarily preserve distance. There are two reasons for doing this. First, we want to be able to recognize and classify figures according to their shapes rather than insisting on the stronger condition of congruence. For example, we want to have transformations that relate similar triangles. The second reason is computational convenience. The algebraic conditions that determine an isometry are more difficult to work with than those based merely on incidence.

Definition. *A* collineation *is a bijection* T: $\mathbf{E}^2 \rightarrow \mathbf{E}^2$ *satisfying the condition that for all triples* P, Q, *and* R *of distinct points,* P, Q, *and* R *are collinear if and only if* TP, TQ, *and* TR *are collinear.*

Although this definition, like the original definition of isometry, is appealing because of its geometric flavor, it does not lend itself immediately to computation. We need a more algebraic version.

Definition. *A mapping* T: $\mathbf{E}^2 \rightarrow \mathbf{E}^2$ *is called an* affine transformation *if there is an invertible 2 by 2 matrix* A *and a vector* $b \in \mathbf{R}^2$ *such that, for all* $x \in \mathbf{R}^2$,

$$Tx = Ax + b.$$

Remark: By Theorem 1.38 every isometry is an affine transformation.

Remark: The matrix A and the vector b mentioned in the definition are uniquely determined by T. In fact, $b = T(0)$ and the columns of A are the vectors $T\varepsilon_i - b$, $i = 1, 2$. We call A the *linear part* of T, and b the *translation part* of T.

Theorem 1. *Every affine transformation is a collineation.*

Proof: The following identity, which can be easily checked, directly shows that affine transformations are surjective.

$$A(A^{-1}(x - b)) + b = x.$$

On the other hand, if $Ax + b = A\bar{x} + b$, then $A(x - \bar{x}) = 0$. Because A is invertible, we must have $x = \bar{x}$. Thus, affine transformations are injective.

Finally, for any points P and Q and any affine transformation T, it is easy to check that

$$T((1 - t)P + tQ) = (1 - t)TP + tTQ \qquad (2.1)$$

for all real t. Thus, if R is a point collinear with P and Q, TR will be collinear with TP and TQ. Conversely, if R' is a point collinear with TP and TQ, there is (because T is surjective) a unique point R with $TR = R'$. But now we know that

$$TR = (1 - t)TP + tTQ \qquad (2.2)$$

for some number t. Because T is injective, (2.1) and (2.2) yield

$$R = (1 - t)P + tQ,$$

and R is collinear with P and Q. □

Corollary. *Every isometry is a collineation.*

Theorem 2. *Every collineation is an affine transformation.*

The proof of Theorem 2 is too technical to present here but is included in Appendix E. From now on in this chapter we will treat the word "collineation" as a synonym for affine transformation.

Fixed lines

If T is an affine transformation and ℓ is a line, then $T\ell$ is a line. We now show how to compute this line in terms of the data determining T and ℓ.

Theorem 3. *Let T be an affine transformation, and let $\ell = P + [v]$ be a line. Then $T\ell$ is the line $TP + [Av]$, where A is the linear part of T.*

Proof: Let b be the translation part of T. For real t,

$$T(P + tv) = A(P + tv) + b = TP + tAv.$$

From this equation we can see that every point of $T\ell$ lies on $TP + [Av]$, and conversely. Note that $T\ell$ is in fact a line because $Av \neq 0$. □

Corollary. *Let T be an affine transformation with linear part A and translation part b. A line $P + [v]$ is a fixed line of T if and only if v is an eigenvector of A and $(A - I)P + b \in [v]$. (The notion of eigenvector is discussed in Appendix D.)*

Theorem 4.
  i.   *If two fixed lines of an affine transformation intersect, they do so in a fixed point.*
 ii.   *If two fixed lines of an affine transformation are parallel, every line in the pencil containing these lines is fixed.*
iii.   *If two lines are parallel, their images under any affine transformation are parallel.*

The reader may prove these facts as an exercise. (See Exercise 2.)
We now have the machinery required to prove Theorem 40 of Chapter 1:

Proof (of Theorem 1.40): Let $\ell = P + [v]$ be a line, and let T be an affine transformation with linear part A and translation part b.

CASE 1: T is a nontrivial translation, so $A = I$ and $b \neq 0$. Then v is automatically an eigenvector of A, and ℓ is a fixed line if and only if $b \in [v]$. Thus, the fixed lines of T are those with direction $[b]$.

CASE 2:   If T is a half-turn about a point C, then from Exercise 1.26, $A = -I$ and $b = 2C$. Again, v is automatically an eigenvector, and ℓ is a fixed line if and only if $-2P + 2C \in [v]$; that is, $C \in P + [v]$. Thus the fixed lines of T are those that pass through C.

Now consider the case of a rotation having $A = \text{rot } \theta \neq \pm I$. Then A has no nonzero eigenvectors (Exercise 1.31); therefore, T can have no fixed lines.

CASE 3: T is a reflection with axis m. Clearly, m is a fixed line. Furthermore, if $m = Q + [w]$, where $|w| = 1$, then for all real t,

$$\Omega_m(Q + tw^\perp) = Q + tw^\perp - 2\langle tw^\perp, w^\perp \rangle w^\perp = Q - tw^\perp.$$

Thus, $Q + [w^\perp]$ is a fixed line. In other words, Ω_m leaves fixed all lines perpendicular to m. By Theorem 4(i), any fixed line not perpendicular to m must meet the pencil of perpendiculars to m in fixed points. Because Ω_m has no fixed points except on m (Theorem 1.21), Ω_m can have no additional fixed lines.

CASE 4: $T = \Omega_m \tau_{kw}$ is a glide reflection consisting of reflection in a line m with unit direction vector w and a translation by a nonzero multiple of w.

We first show that m is a fixed line. Let Q be any point of m. Then for real t,

$$T(Q + tw) = \Omega_m(Q + tw + kw) = Q + (t + k)w - 2\langle(t + k)w, w^\perp\rangle w^\perp$$
$$= Q + (t + k)w.$$

Thus, the line m is a fixed line.

Because T has no fixed points, any other fixed line would have to be parallel to m. Let $\ell = Q + sw^\perp + [w]$ be a typical line parallel to m. Then

$$T(Q + sw^\perp + tw) = \Omega_m(Q + sw^\perp + (t + k)w)$$
$$= Q + sw^\perp + (t + k)w - 2\langle sw^\perp + (t + k)w, w^\perp\rangle w^\perp$$
$$= Q - sw^\perp + (t + k)w.$$

Note that ℓ cannot be fixed unless $s = 0$. $\qquad\qquad\square$

The affine group AF(2)

We now look at the result of successively applying two affine transformations. If

$$Tx = Ax + b \quad \text{and} \quad \tilde{T}x = \tilde{A}x + \tilde{b},$$

then

$$T\tilde{T}x = A(\tilde{A}x + \tilde{b}) + b = (A\tilde{A})x + A\tilde{b} + b.$$

Thus, the composition of two affine transformations is again an affine transformation. One can arrange that $T\tilde{T} = I$ by choosing $\tilde{A} = A^{-1}$ and $\tilde{b} = -A^{-1}b$, thus showing that the inverse of an affine transformation is also an affine transformation. To summarize, we have proved

Theorem 5. *The set* **AF(2)** *of all affine transformations of* $\mathbf{R}^2$ *is a group, called the* affine group *of* $\mathbf{R}^2$.

Elements of **AF(2)** may be conveniently represented by matrices as follows: Write

$$A = \begin{bmatrix} a_{11} & a_{12} \\ a_{21} & a_{22} \end{bmatrix} \quad \text{and} \quad b = \begin{bmatrix} b_1 \\ b_2 \end{bmatrix}.$$

If $y = Ax + b$, we may easily check that the 3 by 3 matrix equation

$$\begin{bmatrix} y_1 \\ y_2 \\ 1 \end{bmatrix} = \begin{bmatrix} a_{11} & a_{12} & b_1 \\ a_{21} & a_{22} & b_2 \\ 0 & 0 & 1 \end{bmatrix} \begin{bmatrix} x_1 \\ x_2 \\ 1 \end{bmatrix} \tag{2.3}$$

holds and that the composition operation in **AF(2)** corresponds to matrix multiplication of the associated 3 by 3 matrices.

If **GL(3)** denotes the group of all invertible 3 by 3 matrices, then **AF(2)** is a subgroup of **GL(3)**. This representation may be abbreviated as

$$T = \begin{bmatrix} A & b \\ 0 & 1 \end{bmatrix},$$

where the sizes of the various matrices are understood from the context. These ideas are formalized in Exercise 6.

Fundamental theorem of affine geometry

Affine geometry consists of those facts about $\mathbf{E}^2$ that depend only on incidence properties and not on perpendicularity or distance. Although affine geometry is interesting in its own right, we will be concentrating here on those aspects that will help us to solve problems of congruence and symmetry of figures.

The fundamental theorem gives a clear and simple criterion for existence and uniqueness of affine transformations, namely, that any two triangles can be related by a unique affine transformation.

At this point it is useful to highlight a fact that arose in the proof of Theorem 1 – affine transformations preserve order along lines.

Theorem 6. *Let P and Q be points, and let T be an affine transformation. Then*
i. *For any real number t,*

$$T((1 - t)P + tQ) = (1 - t)TP + tTQ. \qquad (2.1)$$

ii. *A point X lies between P and Q if and only if TX lies between TP and TQ. Furthermore,*

$$\frac{d(P, X)}{d(P, Q)} = \frac{d(TP, TX)}{d(TP, TQ)}.$$

We are now in a position to derive an important uniqueness property of affine transformations.

Theorem 7.
i. *If an affine transformation leaves fixed two distinct points, then it leaves fixed every point on the line joining these points.*
ii. *If an affine transformation leaves fixed three noncollinear points, it must be the identity.*

Proof: The first claim is immediate from formula (2.1). Now, let P, Q, and R be three noncollinear points that are left fixed by an affine transformation T. Let X be any point not lying on any of the lines $\overleftrightarrow{PQ}$, $\overleftrightarrow{QR}$, or $\overleftrightarrow{RP}$. Let A be the midpoint of the segment PQ. Now $\overleftrightarrow{XA}$ cannot be parallel to both $\overleftrightarrow{QR}$ and $\overleftrightarrow{RP}$; hence, it meets one of these lines in a point B (distinct from A). Because X is on a line containing two fixed points A and B, X itself must be a fixed point. We conclude that T leaves every point in the plane fixed and therefore is the identity. $\square$

Our proof of Theorem 7(ii) is a synthetic proof. It uses geometric ideas derived earlier and geometric arguments.

In following the proof it is helpful to make your own diagram. An alternative proof using linear algebra is suggested in Exercise 7.

We now come to the fundamental theorem, which asserts the existence of a unique affine transformation relating any two triangles.

Theorem 8. *Given two noncollinear triples of points, PQR and $P'Q'R'$, there is a unique affine transformation T such that $TP = P'$, $TQ = Q'$, and $TR = R'$.*

Proof: Because $\{Q - P, R - P\}$ and $\{Q' - P', R' - P'\}$ are bases for $\mathbf{E}^2$ (Appendix D), there is an invertible 2 by 2 matrix A such that $A(Q - P) = Q' - P'$ and $A(R - P) = R' - P'$. Let $T = \tau_{P'}A\tau_{-P}$. Then $TP = \tau_{P'}A(P - P) = P'$. Similarly, $TQ = Q'$ and $TR = R'$. Thus, we have constructed an affine transformation with the required property.

We now show that there is only one such transformation. Suppose that $\tilde{T}$ agrees with T on P, Q, and R. Then $\tilde{T}^{-1}T$ is an affine transformation that leaves P, Q, and R fixed. By Theorem 7, $\tilde{T}^{-1}T = I$; that is, $T = \tilde{T}$. $\square$

Affine Reflections

Let P, Q, and R be noncollinear points of $\mathbf{E}^2$. The unique affine transformation (guaranteed to exist by the fundamental theorem) that leaves P fixed while interchanging Q and R is called an *affine reflection* and is denoted by the symbol (used in [8]).

$$[P; Q \leftrightarrow R]$$

(see Figure 2.1). Clearly, every ordinary reflection is an affine reflection. In fact, let ℓ be any line, P any point on ℓ, Q any point not on ℓ, and $R = \Omega_\ell Q$. Then it is easy to verify (Exercise 8) that

$$\Omega_\ell = [P; Q \leftrightarrow R]. \tag{2.4}$$

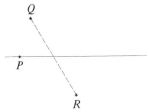

Figure 2.1 The affine reflection $[P; Q \leftrightarrow R]$.

We shall soon see that not every affine reflection is an isometry. However, affine reflections share some of the properties of ordinary reflections. To begin with, an affine reflection T must be involutive because T^2 has three noncollinear fixed points. In addition, we have

Theorem 9. *Let M be the midpoint of a segment QR, and let P be any point not collinear with Q and R. Then the affine reflection $[P; Q \leftrightarrow R]$ leaves fixed every point of $\overleftrightarrow{PM}$ but no other points.*

Proof: We first check that M is a fixed point. To see this, write $Tx = Ax + b$ as usual. Then

$$Q = TR = AR + b \quad \text{and} \quad R = TQ = AQ + b.$$

Hence,

$$Q + R = A(Q + R) + 2b;$$

that is,

$$M = AM + b = TM,$$

and M is a fixed point. By Theorem 7(i), $\overleftrightarrow{PM}$ consists entirely of fixed points. On the other hand, the affine reflection is not the identity, so it cannot have any additional fixed points, by Theorem 7(ii). $\square$

Theorem 10. *The affine reflection $[P; Q \leftrightarrow R]$ leaves fixed the line $\overleftrightarrow{PM}$ and all lines parallel to $\overleftrightarrow{QR}$ and no other lines. (Notation is as for Theorem 9.)*

Proof: Because TQ and TR determine the same line as Q and R, we see that the line $\ell = \overleftrightarrow{QR}$ is fixed. Each line ℓ' parallel to ℓ meets $\overleftrightarrow{PM}$ in a fixed point M'. Thus, $T\ell'$ passes through M' while remaining parallel to $T\ell = \ell$ (Theorems 4(iii) and 1.17). This guarantees that $T\ell' = \ell'$; that is, ℓ' is a fixed line. Finally, suppose that ℓ'' were a fixed line not parallel to ℓ but distinct from $\overleftrightarrow{PM}$. Then ℓ'' meets ℓ' and ℓ in fixed points. This contradicts the fact that all fixed points are on $\overleftrightarrow{PM}$. $\square$

Theorem 11. *The affine reflection $[P; Q \leftrightarrow R]$ is an isometry if and only if $\overleftrightarrow{PM} \perp \overleftrightarrow{QR}$.*

Proof: Suppose that the given affine reflection is an isometry. Because it has a line of fixed points, it must be an ordinary reflection with axis $\overleftrightarrow{PM}$, by Theorem 1.39. But $\overleftrightarrow{QR}$ is a fixed line of this reflection, and thus it must be perpendicular to $\overleftrightarrow{PM}$ by Theorem 1.40.

45

Conversely, suppose that $\overleftrightarrow{PM} \perp \overleftrightarrow{QR}$. Let $m = \overleftrightarrow{PM}$. We show that Ω_m interchanges Q and R and thus, by the fundamental theorem, must coincide with the given affine reflection. To this end, note that

$$\Omega_m Q = Q - 2\langle Q - M, N\rangle N,$$

where N is a unit vector in the direction $[Q - R]$. But

$$Q - M = \tfrac{1}{2}(Q - R).$$

Thus,

$$\Omega_m Q = Q - \langle Q - R, N\rangle N = Q - (Q - R) = R.$$

Also $\Omega_m R = \Omega_m \Omega_m Q = Q$. Hence, Ω_m interchanges Q and R. $\qquad\square$

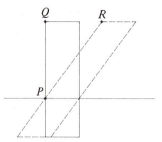

Figure 2.2   The shear $[P; Q \to R]$.

Shears

The fundamental theorem of affine geometry can be used to define other classes of affine transformations. Let P, Q, and R be noncollinear points. The unique affine transformation that leaves fixed every point on the line through P parallel to $\overleftrightarrow{QR}$ and that takes Q to R is denoted by $[P; Q \to R]$ and is called a *shear*. See Figure 2.2.

Theorem 12.   *The shear $[P; Q \to R]$ has the line through P parallel to $\overleftrightarrow{QR}$ as its set of fixed points. The fixed lines are those belonging to the pencil of parallels determined by $\overleftrightarrow{QR}$.*

Proof: Let T be the shear in question. T can have no fixed points other than those on the line $m = P + [Q - R]$. Otherwise, it would be the identity.

Let $\ell = Q + [Q - R]$. Because $\ell \parallel m$ and $Tm = m$, $T\ell$ is the unique line through $R = TQ$ parallel to m. In other words, $T\ell = \ell$, and ℓ is a fixed line.

Finally, let X be any point lying neither on ℓ nor on m, and let $X' = TX$. The line $\overleftrightarrow{XX'}$ must be parallel to m; otherwise $\overleftrightarrow{XX'}$ would have to meet m in a fixed point B and $\overleftrightarrow{XB} = \overleftrightarrow{X'B}$ would be a fixed line. But now the fixed lines $\overleftrightarrow{QR}$ and $\overleftrightarrow{XB}$ would have to intersect in a fixed point, which is impossible. This shows that TX lies on $X + [Q - R]$ and, hence, that $X + [Q - R]$ is a fixed line. Our argument also shows that no other lines of $\mathbf{E}^2$ can be fixed. $\qquad\square$

Remark: The line of fixed points of a shear is called its *axis*.

Theorem 13.   *A shear whose fixed points lie along the x_1-axis has a matrix of the form*

$$s_\lambda = \begin{bmatrix} 1 & \lambda & 0 \\ 0 & 1 & 0 \\ 0 & 0 & 1 \end{bmatrix}. \qquad (2.5)$$

Proof: Because the origin is a fixed point, the translation part is 0. Also, the vector ε_1 is a fixed point, and this determines the first column of the matrix. Finally, the shear must take ε_2 to a point on the horizontal line $\varepsilon_2 + [\varepsilon_1]$. This determines the form of the second column. □

Remark:

i.   Every matrix of the form s_λ, $\lambda \neq 0$, determines a shear.

ii. If T is a shear with axis ℓ, and ρ is any affine transformation, then $\rho T \rho^{-1}$ is a shear with axis $\rho\ell$.

These facts follow easily from the fundamental theorem (Exercise 9). A shear whose axis is a horizontal line through a point P can be written

$$Tx = P + s_\lambda(x - P), \qquad (2.6)$$

and a shear whose axis passes through the origin and has direction vector $(\cos \theta, \sin \theta) = (\text{rot } \theta)\varepsilon_1$ can be written

$$Tx = (\text{rot } \theta)s_\lambda(\text{rot } (-\theta))x,$$

so that any shear with axis $P + [(\text{rot } \theta)\varepsilon_1]$ can be written in the form

$$Tx = P + (\text{rot } \theta)s_\lambda(\text{rot } (-\theta))(x - P) \qquad (2.7)$$

for some real number $\lambda \neq 0$.

Dilatations

A *dilatation* is an affine transformation with the property that for each line ℓ, either $T\ell = \ell$ or $T\ell \parallel \ell$. The identity is said to be a *trivial dilatation*.

Theorem 14.   *A dilatation that leaves two points fixed must be the identity.*

Proof: Suppose that P and Q are distinct fixed points of a dilatation T. Then every point on the line $\overleftrightarrow{PQ}$ is fixed. Let X be any point not on $\overleftrightarrow{PQ}$. Then T takes the line $\overleftrightarrow{PX}$ to a line through P with the same direction. In other words, $\overleftrightarrow{PX}$ is a fixed line. By the same argument $\overleftrightarrow{QX}$ is a fixed line, and so X is a fixed point. Because T has three noncollinear fixed points, it must be the identity.

Thus, a nontrivial dilatation can have at most one fixed point. A dilatation with exactly one fixed point is called a *central dilatation*, and the fixed point is called its *center*. See Figures 2.3 and 2.4.

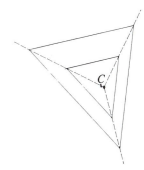

Figure 2.3   Two triangles related by a central dilatation.

47

Theorem 15.

i. *A central dilatation with center C may be written in the form*

$$Tx = C + \kappa(x - C). \qquad (2.8)$$

The number $|\kappa|$ is called the magnification factor of T.

ii. *A dilatation that has no fixed points is a translation.*

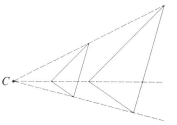

Figure 2.4 Two more triangles related by a central dilatation.

Proof: Let T be a central dilatation with center C. Because T is a dilatation, every vector $v \in \mathbf{R}^2$ must be an eigenvector of A (the linear part of T). Thus, there is a nonzero real number κ such that $A = \kappa I$ (Exercise 10).

Because $TC = C$, the translation part of T is equal to $C - \kappa C$, so that for all $x \in \mathbf{E}^2$,

$$Tx = \kappa x + C - \kappa C = C + \kappa(x - C).$$

This proves (i). Further, if $\kappa \neq 1$, the equation $\kappa x + b = x$ has a solution $x = (-1/(\kappa - 1))b$. Hence, every dilatation is either a translation ($\kappa = 1$) or has a fixed point. □

Theorem 16. *The fixed lines of a central dilatation are precisely those that pass through its center.*

Proof: First, note that

$$T(C + [v]) = TC + [v] = C + [v],$$

so that all lines through C are fixed. On the other hand, if any fixed line ℓ does not pass through C, pick an arbitrary point X on this line. Then $\overleftrightarrow{CX}$ and ℓ are fixed lines intersecting in X. Because X cannot be a fixed point, we have a contradiction. □

Remark: A half-turn is a special central dilatation having $\kappa = -1$.

Similarities

Definition. *A mapping $T: \mathbf{E}^2 \to \mathbf{E}^2$ is called a* similarity *(with magnification factor $\kappa > 0$) if, for all $X, Y \in \mathbf{E}^2$,*

$$d(TX, TY) = \kappa d(X, Y).$$

A similarity can be accomplished in two stages: first, a central dilatation (to make objects the right size); then an isometry (to move objects to the right position).

Theorem 17. *Every similarity is a central dilatation followed by an isometry. In particular, every similarity is an affine transformation.*

Proof: Let T be a similarity with magnification factor κ. Let S be the central dilatation defined by

$$SX = \frac{1}{\kappa}X.$$

Then

$$d(TSX,\ TSY) = \kappa d(SX,\ SY)$$
$$= \kappa\frac{1}{\kappa}d(X,\ Y) = d(X,\ Y).$$

Thus, TS is an isometry, and, hence, T is this same isometry preceded by the central dilatation S^{-1}. □

Theorem 18.

i. *If T_1 and T_2 are similarities with respective magnification factors κ_1 and κ_2, then T_1T_2 is a similarity with magnification factor $\kappa_1\kappa_2$.*

ii. *If T is a similarity with magnification factor κ, then T^{-1} is a similarity with magnification factor $1/\kappa$.*

Proof: Exercise 11. □

Corollary. *The set of similarities of $\mathbf{E}^2$ is a group, which we denote by* $\mathbf{Sim(E^2)}$.

Definition. *Two figures $\mathscr{F}_1$ and $\mathscr{F}_2$ are* similar *if there is a similarity T such that $T\mathscr{F}_1 = \mathscr{F}_2$.*

Affine symmetries

Let $\mathscr{F}$ be any figure. An affine transformation leaving $\mathscr{F}$ fixed is called an *affine symmetry* of $\mathscr{F}$, and the set of all affine symmetries is a group called the *affine symmetry group* of $\mathscr{F}$. We use the notation

$$\mathscr{AS}(\mathscr{F}) = \{T \in \mathbf{AF(2)}|T\mathscr{F} = \mathscr{F}\}.$$

Because every isometry is an affine transformation, we have

$$\mathscr{S}(\mathscr{F}) \subset \mathscr{AS}(\mathscr{F}) \subset \mathbf{AF(2)}.$$

In the next section we will set up a framework for classifying the affine symmetries of a wide class of figures. In the meantime we will examine the symmetries of some very simple figures.

Theorem 19.   *Let $\mathscr{F}$ be the set consisting of a single point. Then $\mathscr{AS}(\mathscr{F}) \cong$
$\mathbf{GL}(2)$, the group of 2 by 2 invertible real matrices.*

Proof: Let P be the point. If $T \in \mathscr{AS}(\mathscr{F})$, then $\tau_{-P}T\tau_P$ leaves 0 fixed, and
its translation part is 0. We write $\tau_{-P}T\tau_P = A$, where A is linear, and
conclude that $T = \tau_P A \tau_{-P}$; that is, $Tx = P + A(x - P)$ for all $x \in \mathbf{E}^2$. It is
now a routine matter to check that the mapping that takes T to A is an
isomorphism (Exercise 12).   □

Theorem 20.   *Let $\mathscr{F}$ be a set consisting of two points. Then $\mathscr{AS}(\mathscr{F})$ is
isomorphic to the group of 2 by 2 matrices of the form*

$$\begin{bmatrix} \pm 1 & \lambda \\ 0 & \mu \end{bmatrix} \quad with \quad \mu \neq 0. \tag{2.9}$$

Writing

$$T^+_{\lambda,\mu} = \begin{bmatrix} 1 & \lambda \\ 0 & \mu \end{bmatrix} \quad and \quad T^-_{\lambda,\mu} = \begin{bmatrix} -1 & \lambda \\ 0 & \mu \end{bmatrix},$$

one can verify that the $T^+_{\lambda,\mu}$ form a subgroup of $\mathscr{AS}(\mathscr{F})$. Some other
interesting subsets are

i.   the subgroup $\mathscr{S}(\mathscr{F})$ determined by $\lambda = 0$, $\mu = \pm 1$,
ii. the subgroup $\{T^+_{\lambda,\mu} | \mu = 1\}$ consisting of all shears leaving the two
     given points fixed together with the identity,
iii. the set $\{T^+_{\lambda,\mu} | \mu = -1\}$ of affine reflections leaving the two points
     fixed,
iv. the subgroup $\{T^+_{\lambda,\mu} | \lambda = 0, \mu > 0\}$ consisting of stretches in the
     direction of the x_2-axis, including the identity as a special case.

Remark: After we have studied projective geometry (Chapter 5), we will
be able to see that these transformations are merely the affine versions of
the projective collineations leaving a line pointwise fixed. In projective
geometry these break down into two types: *homologies* and *elations*.

Rays and angles

Let P be a point of $\mathbf{E}^2$, and let v be a nonzero vector. Then

$$\imath = \{P + tv | t \geqslant 0\} \tag{2.10}$$

is called a *ray* with *origin* P and *direction vector* v. Clearly, every line
through P is the union of two rays with origin P. Their direction vectors are
negatives of each other.
    The union of two rays $\imath_1$ and $\imath_2$ with common origin P is called an *angle*

with vertex P and arms $\imath_1$ and $\imath_2$. We allow the possibility $\imath_1 = \imath_2$, in which case we refer to the angle as a *zero angle*. If $\imath_1$ and $\imath_2$ are two halves of the same line, we say that they are *opposite* rays and that the angle is a *straight angle*. Finally, if $\imath_1 \perp \imath_2$ we call the angle a *right angle*. (Two rays are perpendicular if their direction vectors are orthogonal.)

Given two distinct points P and Q, there is a unique ray with origin P that passes through Q. We denote this ray by $\overrightarrow{PQ}$. The angle with vertex Q and arms $\overrightarrow{QP}$ and $\overrightarrow{QR}$ is denoted by $\measuredangle PQR$ or, equivalently, $\measuredangle RQP$. Rays and angles may be represented as shown in Figures 2.5–2.8.

It is now time to define a numerical measure for angles. This must be done in terms of analytic concepts, taking care not to appeal to our pictorial notions of angle measurement.

Definition. *Let $\mathscr{A}$ be an angle whose arms have unit direction vectors u and v. The* radian measure *of $\mathscr{A}$ is defined to be*

$$\cos^{-1}\langle u, v\rangle. \tag{2.11}$$

Remark: If we write $u = (\cos \theta, \sin \theta)$ and $v = (\cos \phi, \sin \phi)$, then the radian measure α is the unique number in $[0, \pi]$ such that

$$(\text{rot } \alpha)u = v \quad \text{or} \quad (\text{rot } \alpha)v = u.$$

In other words, there is a rotation by α taking one arm of $\mathscr{A}$ to the other. (See Exercise 14.)

Theorem 21. *Let $\mathscr{A}$ be any angle. Its radian measure α is*
  i. *0 if and only if $\mathscr{A}$ is a zero angle,*
 ii. *π if and only if $\mathscr{A}$ is a straight angle,*
iii. *between 0 and π otherwise.*
Furthermore, $\alpha = \pi/2$ if and only if $\mathscr{A}$ is a right angle.

Definition. *An angle $\mathscr{A}$ is* acute *if its radian measure is $<\pi/2$. It is* obtuse *if its radian measure is $>\pi/2$. See Figures 2.9–2.11.*

Theorem 22. *Let $\mathscr{A} = \measuredangle PQR$ be an angle. Then*
  i. *$\measuredangle PQR$ is acute if and only if $\langle P - Q, R - Q\rangle$ is positive.*
 ii. *$\measuredangle PQR$ is obtuse if and only if $\langle P - Q, R - Q\rangle$ is negative.*

Definition. *Let $\mathscr{A}$ be an angle with vertex P and radian measure α. Let u and v be unit direction vectors of its arms chosen so that $(\text{rot } \alpha)u = v$. Then a ray with origin P and direction vector $\text{rot}(\alpha/2)u$ is called a* bisector *of $\mathscr{A}$.*

Remark: A straight angle has two bisectors. Any other angle has a unique bisector.

Figure 2.5　A straight angle.

Figure 2.6　A zero angle.

Figure 2.7　The ray $\overrightarrow{PQ}$.

Figure 2.8　An angle with vertex P.

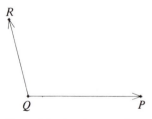

Figure 2.9 An obtuse angle.

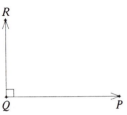

Figure 2.10 A right angle.

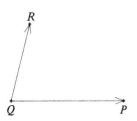

Figure 2.11 An acute angle.

Theorem 23. *For any angle $\mathscr{A}$ there is a unique reflection that interchanges its arms, namely, the reflection in the line containing the bisector(s) of $\mathscr{A}$.*

Proof: Let u, v, and α be as in the previous definition. If P is the vertex of $\mathscr{A}$, set $T = \tau_P(\text{ref}((\theta + \phi)/2))\tau_{-P}$, where $u = (\cos\theta, \sin\theta)$ and $v = (\cos\phi, \sin\phi)$. Then

$$T(P + tu) = P + \left(\text{ref}\left(\frac{\theta + \phi}{2}\right)\right) tu$$

$$= P + t\left(\text{ref}\left(\frac{\theta + \phi}{2}\right)\right)(\text{rot }\theta)\varepsilon_1$$

$$= P + t(\text{rot }\phi)\varepsilon_1$$

$$= P + tv$$

for all real t. Thus, $T\imath_1 = \imath_2$ and, by symmetry, $T\imath_2 = \imath_1$.

To show uniqueness, let $\tilde{T}$ be any other reflection that interchanges $\imath_1$ and $\imath_2$. Then $\tilde{T}T$ leaves the rays $\imath_1$ and $\imath_2$ fixed. Hence, their point of intersection P is fixed. Because $TP = P$, we have that $\tilde{T}P = P$. Thus, the axis of the reflection $\tilde{T}$ passes through P, and $\tilde{T}T$ is a rotation about P. Now the only rotation that leaves a ray fixed is the identity (Exercise 21), and we conclude that $\tilde{T}T$ must be the identity. Thus $T = \tilde{T}$. $\qquad\square$

Rectilinear figures

A union of finitely many segments, rays, and lines is called a *rectilinear figure*. Familiar examples are triangles, squares, and angles. We will study these in detail later. First, we develop some techniques for computing symmetry groups that are applicable to all rectilinear figures.

Let $\mathscr{F}$ be any rectilinear figure. The figure $\hat{\mathscr{F}}$ consisting of all lines that contain lines, segments, or rays of $\mathscr{F}$ is called the *rectilinear completion* of $\mathscr{F}$. A rectilinear figure $\mathscr{F}$ is said to be *complete* if, whenever a segment is in $\mathscr{F}$, the line containing it is in $\mathscr{F}$. Then $\hat{\mathscr{F}}$ is clearly the smallest complete rectilinear figure containing $\mathscr{F}$. See Figures 2.12 and 2.13.

Theorem 24. *Let T be an affine transformation, and let $\mathscr{F}$ be a rectilinear figure. Then T maps the set of lines of $\hat{\mathscr{F}}$ bijectively to the set of lines of $\widehat{T\mathscr{F}}$.*

Proof: We first show that the map is surjective. Suppose that $T\ell$ is a line of $\widehat{T\mathscr{F}}$, but ℓ is not in $\hat{\mathscr{F}}$. Then for each line m of $\hat{\mathscr{F}}$, Tm meets $T\ell$ in at most one point. Because $T\mathscr{F} \cap T\ell$ is contained in the union of all the Tm, it can contain only finitely many points. But this is impossible because $T\ell$ contains at least a segment of $T\mathscr{F}$ and, hence, an infinite number of points of $T\mathscr{F}$.

It only remains to show that if m is any line of $\hat{\mathscr{F}}$, then Tm is a line of $\widehat{T\mathscr{F}}$. First note that m contains a segment m_0 that is contained in $\mathscr{F}$. Then Tm_0 is a segment in $T\mathscr{F}$. Thus, $\widehat{T\mathscr{F}}$ contains the line determined by Tm_0, namely, Tm. $\quad\square$

Corollary. *Suppose T is an affine symmetry of a rectilinear figure $\mathscr{F}$. Then T permutes the lines of its rectilinear completion $\hat{\mathscr{F}}$.*

Definition. *Suppose that $\mathscr{F}$ is a rectilinear figure. A point of $\mathscr{F}$ where two lines of $\hat{\mathscr{F}}$ intersect is called a* vertex *of $\mathscr{F}$.*

Theorem 25. *Let $\mathscr{F}$ be a rectilinear figure and T an affine transformation. Then T maps the set of vertices of $\mathscr{F}$ bijectively to the set of vertices of $T\mathscr{F}$. If T is an affine symmetry of $\mathscr{F}$, then T permutes the vertices of $\mathscr{F}$.*

Figure 2.12 A rectilinear figure.

Proof: We need only show that T maps vertices of $\mathscr{F}$ to vertices of $T\mathscr{F}$. The rest is trivial.

Let P be a vertex of $\mathscr{F}$. Then TP is a point of $T\mathscr{F}$. Because P is the intersection of two lines of $\hat{\mathscr{F}}$, TP is the intersection of their images, which, by Theorem 24, are in $T\mathscr{F}$. Thus, TP is a vertex of $T\mathscr{F}$. $\quad\square$

Corollary.

i. *Every affine symmetry T of a rectilinear figure $\mathscr{F}$ is also an affine symmetry of $\hat{\mathscr{F}}$.*

ii. *Every affine symmetry of a rectilinear figure $\mathscr{F}$ permutes the set of vertices of $\hat{\mathscr{F}}$ that are not vertices of $\mathscr{F}$.*

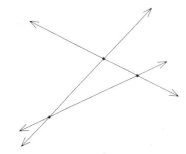

Figure 2.13 Completion of the rectilinear figure in Figure 2.12.

Proof:

i. Because T permutes the lines of $\hat{\mathscr{F}}$ and $\hat{\mathscr{F}}$ is the union of these lines, we must have $T\hat{\mathscr{F}} = \hat{\mathscr{F}}$.

ii. Because T is an affine symmetry of $\hat{\mathscr{F}}$, T must permute the set of vertices of $\hat{\mathscr{F}}$. But T being an affine symmetry of $\mathscr{F}$ must permute the vertices of $\mathscr{F}$ among themselves (i.e., this set is invariant under the permutation). Hence, T must permute the remaining vertices of $\hat{\mathscr{F}}$ among themselves. $\quad\square$

Corollary. *If $\mathscr{F}$ is a rectilinear figure having at least three noncollinear vertices, then $\mathscr{A}\mathscr{S}(\mathscr{F})$ is a finite group.*

Proof: Denote the three noncollinear vertices by P, Q, and R. Then each permutation of the vertices of $\mathscr{F}$ can be realized by *at most* one affine transformation. The only possible candidate is the unique affine transformation (guaranteed by the fundamental theorem) that agrees with the given permutation on P, Q, and R. In general, of course, this candidate may fail to agree with the permutation on some other vertex. Even if it accomplishes the permutation of the vertices, it may fail to be a symmetry of $\mathscr{F}$. In any case, the affine symmetry group has at most $n!$ elements, where n is the number of vertices. □

Remark: If all the vertices of a rectilinear figure $\mathscr{F}$ are collinear, then $\mathscr{AS}(\mathscr{F})$ is still a finite group, but the preceding proof will not work. Complete figures of this form can be described explicitly, and we shall explore these in Chapter 3, Exercise 7.

The centroid

Let F be a finite set of points of $\mathbf{E}^2$. For $x \in \mathbf{E}^2$ define

$$f(x) = \sum_{P \in F} d(x, P)^2. \qquad (2.12)$$

Theorem 26. *There is a unique point of $\mathbf{E}^2$ where the function f achieves its minimum value. This point is called the* centroid *of F.*

Proof: Let n be the number of points of F. Then

$$f(x) = \sum_P \langle x - P, x - P \rangle$$

$$= \sum_P (|x|^2 - 2\langle x, P \rangle + |P|^2)$$

$$= n|x|^2 - 2\langle x, \Sigma P \rangle + \Sigma|P|^2.$$

Write $C = (1/n)\Sigma P$ and $b = (1/n)\Sigma|P|^2$.

Then

$$f(x) = n(|x|^2 - 2\langle x, C \rangle + b)$$
$$= n(|x|^2 - 2\langle x, C \rangle + |C|^2 + b - |C|^2)$$
$$= n|x - C|^2 + n(b - |C|^2).$$

Clearly, $f(x)$ is minimum precisely when $x = C$. □

Remark: If $\mathscr{F}$ is a rectilinear figure with a finite number of vertices, the centroid of the set of vertices is often referred to as the centroid of $\mathscr{F}$.

Theorem 27. *Suppose that $\mathscr{F}$ is a rectilinear figure having a finite nonzero number of vertices. Let C be the centroid of $\mathscr{F}$. Then, for any isometry T, TC is the centroid of $T\mathscr{F}$.*

Proof: First, note that P is a vertex of $\mathscr{F}$ if and only if TP is a vertex of $T\mathscr{F}$. Also, the quantity

$$\sum_P |x - TP|^2 = \sum |TT^{-1}x - TP|^2 = \sum |T^{-1}x - P|^2$$

has its minimum value when $T^{-1}x$ is the centroid C of $\mathcal{F}$; that is, $x = TC$. Thus, TC is the centroid of $T\mathcal{F}$. $\qquad\qquad\qquad\qquad\qquad\qquad\square$

Corollary. *If T is a symmetry of a rectilinear figure $\mathcal{F}$ with a finite number of vertices, then T leaves the centroid of $\mathcal{F}$ fixed.*

Symmetries of a segment

Let PQ be a segment. We compute its symmetry group $\mathcal{S}(PQ)$. First of all, we know that any affine transformation T takes the segment PQ to a segment $P'Q'$, where $P' = TP$ and $Q' = TQ$. We first show that T permutes the set $\{P, Q\}$.

Lemma. *A segment determines its end points; that is, if PQ and $\tilde{P}\tilde{Q}$ denote the same segment, then $\{P, Q\} = \{\tilde{P}, \tilde{Q}\}$.*

Proof: Interchanging $\tilde{P}$ and $\tilde{Q}$ if necessary, we may write

$$\tilde{P} = (1 - t)P + tQ \quad \text{and} \quad \tilde{Q} = (1 - s)P + sQ,$$

where $0 \leqslant t < s \leqslant 1$. Now, there exist $\tilde{t}$ and $\tilde{s}$ in $[0, 1]$ such that

$$\begin{aligned}
P &= (1 - \tilde{t})\tilde{P} + \tilde{t}\tilde{Q} \\
&= (1 - \tilde{t})((1 - t)P + tQ) + \tilde{t}((1 - s)P + sQ) \\
&= P + (t + \tilde{t}(s - t))(Q - P).
\end{aligned}$$

Because $Q \neq P$, we must have $t + \tilde{t}(s - t) = 0$. But the conditions $t \geqslant 0$, $\tilde{t} \geqslant 0$, $s - t > 0$ imply that $t = 0$ and $\tilde{t}(s - t) = 0$; that is, $\tilde{t} = 0$. This proves that $\tilde{P} = P$. The proof that $Q = \tilde{Q}$ is similar. $\qquad\qquad\square$

There are two isometries that leave $\{P, Q\}$ pointwise fixed: the reflection Ω_ℓ with axis $\ell = \overleftrightarrow{PQ}$, and the identity. On the other hand, there are exactly two isometries that interchange P and Q. Clearly, one is the reflection Ω_m whose axis m is the perpendicular bisector of PQ. But if T is any other isometry interchanging P and Q, the composition $\Omega_m T$ leaves P and Q fixed. This gives

$$\Omega_m T = I \quad \text{and} \quad T = \Omega_m,$$

or

$$\Omega_m T = \Omega_\ell \quad \text{and} \quad T = \Omega_\ell \Omega_m = H_M,$$

where M is the midpoint of PQ. Thus, $\mathcal{S}(PQ)$ consists of four elements: two reflections, a half-turn, and the identity. The multiplication table for this group is

	I	Ω_ℓ	Ω_m	H_M
I	I	Ω_ℓ	Ω_m	H_M
Ω_ℓ	Ω_ℓ	I	H_M	Ω_m
Ω_m	Ω_m	H_M	I	Ω_ℓ
H_M	H_M	Ω_m	Ω_ℓ	I

The abstract group having this multiplication table is called the Klein four-group.

We state the results we have outlined as a theorem. You will be asked to fill in the details of the proof in Exercise 23.

Theorem 28. *The symmetry group of a segment has four elements: two reflections, a half-turn, and the identity. The group multiplication table is as indicated beforehand.*

Symmetries of an angle

Let $\mathcal{A}$ be an angle other than a straight angle. In this section we compute the symmetry group $\mathcal{S}(\mathcal{A})$.

We first prove a uniqueness lemma.

Lemma. *Let $\mathcal{A}$ be an angle with vertex P. Suppose that an affine symmetry T of $\mathcal{A}$ leaves both lines of $\hat{\mathcal{A}}$ fixed. Then T leaves both arms of $\mathcal{A}$ fixed.*

Proof: Let ℓ_1 and ℓ_2 be the lines and $\imath_1$ and $\imath_2$ the associated rays. Let v and w be unit direction vectors of $\imath_1$ and $\imath_2$, respectively. Write $Tx = Ax + b$. Because $TP = P$, we get $AP + b = P$, so that we may write

$$Tx = A(x - P) + P.$$

Because ℓ_1 is a fixed line, $[Av] = [v]$ (Theorem 3). Thus, there is a real number λ such that $Av = \lambda v$. Now $T(P + v) = Av + P = P + \lambda v$. Because T maps $\mathcal{A}$ into $\mathcal{A}$, $P + \lambda v$ must be in $\mathcal{A}$. Hence, λ is positive, and $T\imath_1 = \imath_1$. Similarly, $T\imath_2 = \imath_2$. $\qquad\square$

Corollary. *In the lemma if T is an isometry, then T is the identity.*

Proof: The string of equalities

$$1 = |v| = d(P + v, P) = d(T(P + v), TP) = d(P + \lambda v, P) = \lambda|v| = \lambda$$

yields $Av = v$. By symmetry, $Aw = w$, and, hence, A is the identity matrix. Finally, for all x,

$$Tx = A(x - P) + P = x - P + P = x,$$

so that T is the identity. $\qquad\qquad\qquad\qquad\qquad\qquad\qquad\square$

Remark: If $\mathscr{A}$ is a straight angle, then (as a set of points) $\mathscr{A}$ is just a line. If $\mathscr{A}$ is a zero angle, then $\mathscr{A}$ is a ray.

Theorem 29. *Suppose that T is an affine symmetry of an angle $\mathscr{A}$. Then T permutes the arms of $\mathscr{A}$.*

Remark: In case $\mathscr{A}$ is a zero angle, we interpret this to mean that T leaves the one arm of $\mathscr{A}$ fixed.

Proof: Let $\imath_1$ and $\imath_2$ be the arms. Let ℓ_1 and ℓ_2 be the lines containing $\imath_1$ and $\imath_2$, respectively. Let m be the line that contains the bisector of $\mathscr{A}$. By the corollary to Theorem 23, T permutes the lines ℓ_1 and ℓ_2. If $T\ell_1 = \ell_1$ and $T\ell_2 = \ell_2$, then the lemma implies that T leaves fixed $\imath_1$ and $\imath_2$. If $T\ell_1 = \ell_2$ and $T\ell_2 = \ell_1$, then $\Omega_m T$ leaves fixed ℓ_1 and ℓ_2 and, hence, $\imath_1$ and $\imath_2$. But then

$$T\imath_1 = \Omega_m \Omega_m T\imath_1 = \Omega_m \imath_1 = \imath_2$$

and

$$T\imath_2 = \Omega_m \Omega_m T\imath_2 = \Omega_m \imath_2 = \imath_1. \qquad\qquad \square$$

Corollary. *$\mathscr{S}(\mathscr{A})$ consists of two elements Ω_m and I.*

Proof: If $T \in \mathscr{S}(\mathscr{A})$, then either T leaves $\imath_1$ and $\imath_2$ fixed and is therefore the identity, or $\Omega_m T$ leaves $\imath_1$ and $\imath_2$ fixed and is therefore the identity. In the latter case, $T = \Omega_m$. $\qquad\qquad\qquad\qquad\qquad\qquad\square$

Remark: This also shows that if $\imath$ is any ray, then $\mathscr{S}(\imath)$ consists of two elements: reflection in the line of $\imath$, and the identity.

The following will be useful when we discuss triangles.

Theorem 30. *Let $\angle PQR$ be an angle. Suppose $d(P, Q) = d(Q, R)$. Let m be the line containing the bisector of $\angle PQR$. Then Ω_m interchanges P and R while leaving Q fixed.*

Proof: Write $P - Q = |P - Q|u$ and $R - Q = |R - Q|v$, and use the notation of Theorem 23. It is sufficient to check that $\Omega_m P = R$.

$$\Omega_m P = \tau_Q \left(\text{ref}\left(\frac{\theta + \phi}{2}\right) \right) \tau_{-Q} P$$

$$= \tau_P\left(\mathrm{ref}\left(\frac{\theta + \phi}{2}\right)\right)|P - Q|u$$

$$= Q + |P - Q|v = Q + |R - Q|v$$

$$= Q + R - Q = R. \qquad \square$$

Barycentric coordinates

Let P, Q, and R be noncollinear points. For each point $X \in \mathbf{E}^2$ there is a unique triple (λ, μ, ν) of real numbers such that

$$X = \lambda P + \mu Q + \nu R \qquad (2.13)$$

and $\lambda + \mu + \nu = 1$. The association

$$X \to \begin{bmatrix} \lambda \\ \mu \\ \nu \end{bmatrix}$$

is called a *barycentric* coordinate system, and PQR is called the triangle of reference. (See Exercise 25.)

Remark:
  i. This generalizes the fact that points on a line $\overleftrightarrow{PQ}$ may be uniquely written as $\lambda P + \mu Q$, where $\lambda + \mu = 1$.
  ii. We will see that the values of the barycentric coordinates λ, μ, and ν relate in a nice way to the position of X with respect to the triangle of reference.
  iii. Barycentric coordinates have a physical interpretation. If weights of λ, μ, and ν are placed at P, Q, and R, respectively, the center of mass of the resulting configuration will be at X. This also applies, of course, to the case of a line, as described in (i).

Theorem 31. *Let $\overleftrightarrow{PQ}$ be a line, and let R be any point not on $\overleftrightarrow{PQ}$. Using PQR as a triangle of reference, we have, for any point X with barycentric coordinates λ, μ, ν,*
  i. *$\nu = 0$ if and only if X lies on $\overleftrightarrow{PQ}$.*
  ii. *$\nu > 0$ if and only if $XR \cap \overleftrightarrow{PQ} = \varnothing$.*

Proof: If $\nu = 0$, this means that

$$X = \lambda P + \mu Q = (1 - \mu)P + \mu Q, \qquad (2.14)$$

and, hence, X lies on $\overleftrightarrow{PQ}$. Conversely, if X lies on $\overleftrightarrow{PQ}$, then (2.14) holds for some value of μ, and by uniqueness of the representation in (2.13), we must have $\nu = 0$.

If $v > 0$, then, for $0 \leq t \leq 1$,

$$(1 - t)X + tR = (1 - t)\lambda P + (1 - t)\mu Q + (1 - t)vR + tR.$$

Because $(1 - t)v + t > 0$, XR cannot intersect $\overleftrightarrow{PQ}$. On the other hand, if $v < 0$, there is a value of t satisfying $(1 - t)v + t = 0$; that is,

$$t = \frac{-v}{1 - v}.$$

Note that $0 < -v/(1 - v) < 1$. Thus, $\overleftrightarrow{PQ} \cap XR \neq \varnothing$. □

Definition. *Let ℓ be a line, and let R be a point not on ℓ. The* half-plane *determined by ℓ and R is the set of X such that $XR \cap \ell = \varnothing$. See Figure 2.14.*

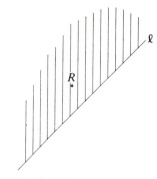

Figure 2.14 A half-plane consisting of all points X such that $XR \cap \ell = \varnothing$.

Theorem 32.

i.   *Every line ℓ determines two half-planes. The reflection Ω_ℓ interchanges the half-planes.*
ii. *Let ℓ be a line, and let P and Q be arbitrary points on ℓ. Let R be any point not on ℓ. Then the half-plane determined by ℓ and R is the set of points having $v > 0$. (Again the triangle of reference is PQR.) The set of points having $v < 0$ is the half-plane determined by ℓ and $\Omega_\ell R$. The two half-planes are said to be* opposites *of each other.*

Remark: When two points are in the same half-plane, we say that they are *on the same side* of ℓ. Points in opposite half-planes are said to be *on opposite sides* of ℓ.

Definition. *A point X is said to be in the* interior *of an angle $\measuredangle PQR$ if $\lambda > 0$ and $v > 0$. See Figure 2.15.*

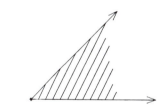

Figure 2.15 The interior of an angle.

Remark: This is the same as saying that X and R are on the same side of $\overleftrightarrow{PQ}$ while X and P are on the same side of $\overleftrightarrow{QR}$.

Theorem 33 (The crossbar theorem). *Let X be a point in the interior of the angle $\measuredangle PQR$. Then the ray $\overrightarrow{QX}$ intersects the segment PR. (See Figure 2.16.)*

Proof: Using PQR as triangle of reference, we obtain

$$Q + t(X - Q) = (1 - t)Q + t\lambda P + t\mu Q + tvR$$
$$= t\lambda P + (1 - t + t\mu)Q + tvR.$$

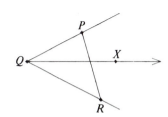

Figure 2.16 The crossbar theorem.

We need to choose a positive value of t so that $1 - t + t\mu = 0$; that is,

$$t = \frac{-1}{\mu - 1} = \frac{1}{1 - \mu}.$$

59

Because $1 - \mu = \lambda + \nu > 0$, this value of t is indeed positive. Furthermore, $\lambda t = \lambda/(1 - \mu) > 0$ and $\nu t = \nu/(1 - \mu) > 0$, so that $Q + t(X - Q)$ lies on PR. $\square$

Addition of angles

Theorem 34.

i. *Let $\angle PQR$ be an angle and X a point in its interior. Then the radian measure of $\angle PQR$ is the sum of the radian measures of $\angle PQX$ and $\angle RQX$. (See Figure 2.17.)*

ii. *Let $\angle PQR$ be a straight angle and X any point not on the line $\overleftrightarrow{PQ}$. Then the sum of the radian measures of $\angle PQX$ and $\angle RQX$ is equal to π.*

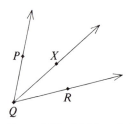

Figure 2.17 Addition of $\angle PQX$ and $\angle XQR$ to form $\angle PQR$.

Remark: In (ii), $\angle PQX$ and $\angle RQX$ are said to be *supplements* of each other. We speak of the pair as a pair of supplementary angles.

Proof: (i) Let θ, θ_1, and θ_2 be the respective radian measures. There is no loss of generality in assuming that $u = P - Q$, $v = R - Q$, and $w = X - Q$ are unit vectors and that $(\text{rot } \theta)u = v$. Because X is in the interior of $\angle PQR$, we may write

$$X - Q = \lambda(P - Q) + \mu(R - Q);$$

that is,

$$w = \lambda u + \mu v,$$

where λ and μ are positive. According to the definition of radian measure, there are four possibilities:

1. $(\text{rot } \theta_1)u = w$ and $(\text{rot } \theta_2)w = v$. Then $(\text{rot}(\theta_1 + \theta_2))u = v$ and $\theta_1 + \theta_2 \equiv \theta \pmod{2\pi}$. But $0 < \theta_1 + \theta_2 < 2\pi$, so that, in fact, $\theta_1 + \theta_2 = \theta$, as required.

    The other three possibilities cannot occur. We examine them in turn.

2. $(\text{rot } \theta_1)u = w$ and $(\text{rot } \theta_2)v = w$. Then

$$0 < \sin \theta_1 = \langle u^{\perp}, w \rangle = \mu \langle u^{\perp}, v \rangle$$

and

$$0 < \sin \theta_2 = \langle v^{\perp}, w \rangle = \lambda \langle v^{\perp}, u \rangle.$$

    But $\langle u^{\perp}, v \rangle = \langle u^{\perp\perp}, v^{\perp} \rangle = -\langle u, v^{\perp} \rangle$, so we have a contradiction.

3. $(\text{rot } \theta_1)w = u$ and $(\text{rot } \theta_2)w = v$. This is similar to case (2). We get the same expressions for the negative numbers $\sin(-\theta_1)$ and $\sin(-\theta_2)$ and, thus, a contradiction.

4. $(\mathrm{rot}\ \theta_1)w = u$ and $(\mathrm{rot}\ \theta_2)v = w$. In this case,

$$0 < \sin \theta_2 = \lambda \langle u, v^\perp \rangle \quad \text{as in (2)}.$$

But

$$0 < \sin \theta = \langle u^\perp, v \rangle = -\langle u, v^\perp \rangle,$$

a contradiction.

For part (ii) with $\measuredangle PQR$ a straight angle, we have no expression for w in terms of u and v. But $v = -u$ and $v^\perp = -u^\perp$. This makes the proof easier. For (1), $\theta = \pi$ and $\theta_1 + \theta_2 = \pi$ by the same argument. For (2),

$$0 < \sin \theta_1 = \langle u^\perp, w \rangle = -\langle v^\perp, w \rangle = -\sin \theta_2 < 0,$$

a contradiction. Case (3) is similar. Case (4) can occur and gives

$$\theta_1 + \theta_2 = \pi$$

as in (1). $\qquad\qquad\qquad\qquad\qquad\qquad\qquad\qquad\qquad\qquad\square$

Triangles

Let P, Q, and R be noncollinear points of $\mathbf{E}^2$. The *triangle PQR* (sometimes written $\triangle PQR$) is the rectilinear figure consisting of the segments PQ, QR, and PR. These segments are called the *sides* of the triangle.

Theorem 35. *Let PQR be a triangle. Using PQR as the triangle of reference for a barycentric coordinate system, we have that*
  i. *A point $X \in \mathbf{E}^2$ is a vertex of $\triangle PQR$ if and only if exactly two barycentric coordinates are zero.*
  ii. *A point X is on the figure $\triangle PQR$ if it is a vertex or if one barycentric coordinate is zero and the others are positive.*

Definition. *A point is in the* interior *of $\triangle PQR$ if it is in the interior of all three angles determined by P, Q, and R.*

Remark: Points in the interior of the triangle are characterized by having all three barycentric coordinates positive. Figure 2.18 shows the whole plane divided into seven regions characterized by the signs of the barycentric coordinates λ, μ, ν. For example, the interior of the triangle is characterized by the combination $+++$.

Theorem 36. *An affine transformation T takes a triangle $\triangle PQR$ to the triangle $\triangle P'Q'R'$, where $P' = TP$, $Q' = TQ$, and $R' = TR$.*

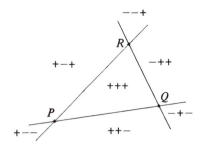

Figure 2.18 Regions of the plane as characterized by the signs of the barycentric coordinates.

Proof: $\triangle PQR$ is the union of three segments belonging to three distinct lines. According to Theorem 6, these segments are transformed by T to the respective segments making up $\triangle P'Q'R'$. The fact that P', Q', and R' are noncollinear and thus form a triangle relies on knowing that T^{-1} is an affine transformation and thus preserves collinearity (Theorems 1 and 6).

$\square$

Symmetries of a triangle

Let Δ be a triangle. If T is an affine symmetry of Δ, then T permutes the vertices of Δ. Conversely, by the fundamental theorem, every permutation of the vertices is realized by a unique affine transformation. Thus $\mathscr{AS}(\Delta)$ is the group of six elements known as the symmetric group $\mathbf{S}_3$. Algebraically, we may describe the group as $\{I, \alpha, \alpha^2, \beta, \alpha\beta, \alpha^2\beta\}$, where $\beta\alpha = \alpha^2\beta$ and $\alpha^3 = I$. In terms of permutations we can set

$$\alpha = (PQR), \quad \beta = (PQ).$$

Clearly, β is the affine reflection $[R; P \leftrightarrow Q]$. Note also that the product of the two affine reflections

$$[R; P \leftrightarrow Q], \quad [P; Q \leftrightarrow R]$$

corresponds to the permutation $(PQR) = \alpha$. Here is the multiplication table for the group $\mathbf{S}_3$:

	I	α	α^2	β	$\alpha\beta$	$\alpha^2\beta$
I	I	α	α^2	β	$\alpha\beta$	$\alpha^2\beta$
α	α	α^2	I	$\alpha\beta$	$\alpha^2\beta$	β
α^2	α^2	I	α	$\alpha^2\beta$	β	$\alpha\beta$
β	β	$\alpha^2\beta$	$\alpha\beta$	I	α^2	α
$\alpha\beta$	$\alpha\beta$	β	$\alpha^2\beta$	α	I	α^2
$\alpha^2\beta$	$\alpha^2\beta$	$\alpha\beta$	β	α^2	α	I

We now investigate $\mathscr{S}(\Delta)$. Clearly, $\mathscr{S}(\Delta)$ is a subgroup of $\mathscr{AS}(\Delta)$.

Definition. *A triangle is*
i.   scalene *if all three sides have different lengths;*
ii. isosceles *if exactly two sides have equal lengths;*
iii. equilateral *if all three sides have the same length.*

Theorem 37. $\mathscr{S}(\Delta)$ *consists of*
i.   *The identity only if Δ is scalene.*

ii. $\{I, \Omega\}$ if Δ is isosceles with $d(P, Q) = d(P, R)$. Ω is the affine reflection $[P; Q \leftrightarrow R]$. Of course, Ω is an actual reflection (isometry) in this case.

iii. All elements of $\mathscr{AS}(\Delta)$ if Δ is equilateral. In this case, two elements are nontrivial rotations about the centroid, three are reflections (one in each median), and the sixth is the identity. (A median is a line that passes through a vertex and the midpoint of the opposite side.)

Proof: Because we already know the affine symmetries, it is only necessary to check which of these are isometries. If $T = [P; Q \leftrightarrow R]$ is an isometry, we must have $d(P, Q) = d(TP, TQ) = d(P, R)$, so that at least two sides of Δ must be of equal length. The same holds for the other two affine reflections.

If T is an isometry that permutes the vertices cyclically, then

$$d(P, Q) = d(TP, TQ) = d(Q, R), \quad \text{say,}$$
$$= d(TQ, TR) = d(R, P),$$

so that Δ must be equilateral.

Thus, when Δ is scalene, only the identity can be an isometry. If Δ is isosceles, then $[P; Q \leftrightarrow R]$ is an isometry (Theorem 11 and Exercise 8). Finally, if Δ is equilateral, all three affine reflections are isometries. The cyclic permutations, being products of reflections, are ordinary rotations. $\qquad \square$

Corollary. Let Δ be an equilateral triangle with centroid C. Then $\mathscr{S}(\Delta)$ consists of

i. the identity,

ii. the three reflections in the medians of Δ,

iii. rotations by $\pm 2\pi/3$ about C.

Proof: First note that

$$C = \tfrac{1}{3}(P + Q + R) = \tfrac{1}{3}P + \tfrac{2}{3}(\tfrac{1}{2}Q + \tfrac{1}{2}R). \qquad (2.15)$$

Similarly,

$$C = \tfrac{1}{3}Q + \tfrac{2}{3}(\tfrac{1}{2}P + \tfrac{1}{2}R)$$

and

$$C = \tfrac{1}{3}R + \tfrac{2}{3}(\tfrac{1}{2}P + \tfrac{1}{2}Q).$$

This exhibits C as a point on all three medians. Thus, the product of two reflections in $\mathscr{S}(\Delta)$ is a rotation about C. In fact, if we write

$$T = \tau_M(\text{rot } \theta)\tau_{-\dot{M}},$$

then

$$T^2 = \tau_M(\text{rot } 2\theta)\tau_{-M} \quad \text{and} \quad T^3 = \tau_M(\text{rot } 3\theta)\tau_{-M}.$$

Now $T^3 = I$ if and only if rot θ = rot $(\pm 2\pi/3)$. $\square$

We have also deduced the following well-known property of the centroid.

Corollary. *For any triangle the centroid lies on each median and divides it in the ratio of $2:1$.*

Congruence of angles

Theorem 38. *Two angles are congruent if and only if they have the same radian measure.*

Proof: Let $\mathcal{A}$ and $\mathcal{B}$ be congruent angles, and let T be an isometry such that $T\mathcal{A} = \mathcal{B}$. By Theorem 24, T maps the two lines of $\hat{\mathcal{A}}$ to the two lines of $\hat{\mathcal{B}}$ and, hence, the vertex of $\mathcal{A}$ to the vertex of $\mathcal{B}$. Let u and v be unit direction vectors for the arms of $\mathcal{A}$. If A is the linear part of T, the arms of $\mathcal{B}$ must have Au and Av as direction vectors. Because A is orthogonal, that is, $\langle Au, Av \rangle = \langle u, v \rangle$, the two angles have the same radian measure.

Conversely, suppose that angles $\mathcal{A}$ and $\mathcal{B}$ have the same radian measure. We may assume that (rot θ)$u = v$ and (rot θ)$u' = v'$, where the arms of $\mathcal{A}$ (respectively, $\mathcal{B}$) have unit direction vectors u, v (respectively, u', v'). Let ϕ be a number such that

$$u' = (\cos \phi)u + (\sin \phi)u^\perp = (\text{rot } \phi)u.$$

Then

$$v' = (\text{rot } \theta)u' = (\text{rot}(\theta + \phi))u$$
$$= (\text{rot } \phi)(\text{rot } \theta)u = (\text{rot } \phi)v.$$

Let P and Q be the respective vertices of $\mathcal{A}$ and $\mathcal{B}$. Then for $t \geqslant 0$,

$$\tau_Q(\text{rot } \phi)\tau_{-P}(P + tu) = \tau_Q(\text{rot } \phi)tu$$
$$= \tau_Q(tu') = Q + (tu')$$

Similarly,

$$\tau_Q(\text{rot } \phi)\tau_{-P}(P + tv) = Q + tv'.$$

Thus, $\mathcal{A}$ and $\mathcal{B}$ are congruent. $\square$

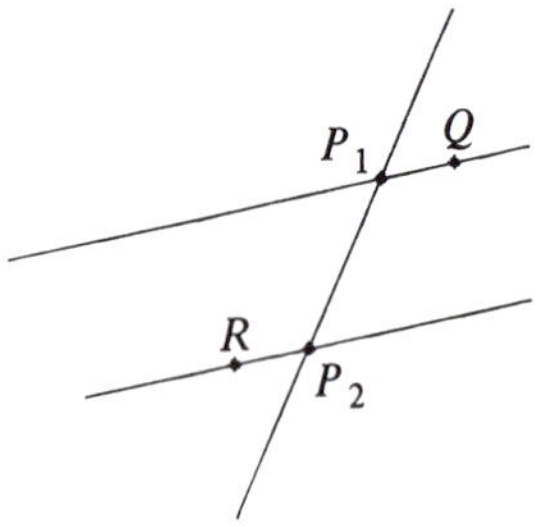

Figure 2.19 A transversal determines two pairs of alternate angles.

A line that intersects two lines in distinct points is called a *transversal* to these lines. Let ℓ_1 and ℓ_2 be parallel lines with direction vector v. Suppose that m is a transversal meeting ℓ_1 and ℓ_2 in P_1 and P_2, respectively. Let $Q = P_1 + v$ and $R = P_2 - v$. Then $\angle P_2P_1Q$ and $\angle RP_2P_1$ are called *alternate* angles. Note that a transversal determines two pairs of alternate angles. (See Figure 2.19.)

Theorem 39. *When a transversal meets two parallel lines, the pairs of alternate angles they determine are congruent.*

Proof: Use the notation introduced in the paragraph preceding Theorem 39. Note that $H_{P_2}\tau_{P_2-P_1}$ takes $\overrightarrow{P_1Q}$ to $\overrightarrow{P_2R}$ and $\overrightarrow{P_1P_2}$ to $\overrightarrow{P_2P_1}$. $\qquad\square$

Remark: The isometry $H_{P_2}\tau_{P_2-P_1}$ is in fact a half-turn about the midpoint of P_1P_2.

Congruence theorems for triangles

We now prove the well-known congruence theorems of Euclidean geometry. The first one (sometimes referred to as the SSS theorem) says that two triangles whose vertices can be matched in such a way that corresponding sides have equal length must be congruent.

Theorem 40. *Let $\triangle PQR$ and $\triangle P'Q'R'$ be such that $d(P, Q) = d(P', Q')$, $d(P, R) = d(P', R')$, and $d(Q, R) = d(Q', R')$. See Figure 2.20. Then there is an isometry T such that $TP = P'$, $TQ = Q'$, and $TR = R'$.*

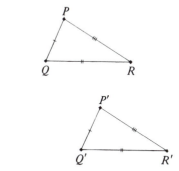

Figure 2.20 The SSS theorem.

Proof: Our method of proof is like that of Euclid, making use of isometries to carry out the "superposition" on which the ancient proof relies. The proof is divided into four steps.

1. Given two lines ℓ_1, ℓ_2, there is an isometry T such that $T\ell_1 = \ell_2$ (i.e., $\mathcal{I}(\mathbf{E}^2)$ is transitive on the lines of $\mathbf{E}^2$). To see this, note that if $\ell_1 \| \ell_2$, reflection in the line lying halfway between them will interchange ℓ_1 and ℓ_2. On the other hand, if ℓ_1 intersects ℓ_2, a reflection in any of the bisectors of the angles they form will interchange ℓ_1 and ℓ_2.
2. If PQ and $P'Q'$ are collinear segments of equal length, there is an isometry T such that $TP = P'$ and $TQ = Q'$. This can be done by $\tau_{P'-P}$ or $H_{P'}\tau_{P'-P}$ (Exercise 32).
3. Suppose that $d(P, R) = d(P, R')$ and $d(Q, R) = d(Q, R')$. Then there is an isometry leaving $\ell = \overleftrightarrow{PQ}$ pointwise fixed and taking R to R'. In fact, I or Ω_ℓ will do (Exercise 33).
4. Choose an isometry T_1 taking $\overleftrightarrow{PQ}$ to $\overleftrightarrow{P'Q'}$. Then choose T_2 to take T_1P to P' and T_1Q to Q'. Let T_3 map T_2T_1R to R' while leaving $\overleftrightarrow{P'Q'}$ pointwise fixed. Then $T = T_3T_2T_1$ accomplishes the desired effect. $\quad\square$

Another famous assertion of Euclid proves congruence based on lengths of two sides and the angle they determine (the SAS theorem). In order to prove this, we first derive the so-called Law of Cosines.

Lemma. Let P, Q, and R be three points of $\mathbf{E}^2$. Then

$$d(P, R)^2 = d(P, Q)^2 + d(Q, R)^2 - 2d(P, Q)\, d(Q, R) \cos \theta, \quad (2.16)$$

where θ is the radian measure of $\sphericalangle PQR$.

Proof: Apply the polarization identity (Exercise 1.29) with $x = P - Q$ and $y = R - Q$, so that $x - y = P - R$. Also, recall that

$$\cos \theta = \frac{\langle P - Q, R - Q \rangle}{|P - Q||R - Q|}, \quad (2.17)$$

from (2.11). □

Theorem 41. *Let $\triangle PQR$ and $\triangle P'Q'R'$ be such that $d(P, Q) = d(P', Q')$, $d(Q, R) = d(Q', R')$, and $\sphericalangle PQR = \sphericalangle P'Q'R'$ (in radian measure). Then there is an isometry T such that $TP = P'$, $TQ = Q'$, and $TR = R'$.*

Proof: Apply the Law of Cosines to both triangles. The given conditions say that the right sides of (2.16) are equal. Hence, $d(P, R) = d(P', R')$, and the SSS theorem may be applied. □

Corollary. *The base angles of an isosceles triangle are congruent.*

Proof: Apply the SAS theorem to $\triangle PQR$ and $\triangle RQP$, where $d(P, Q) = d(Q, R)$. □

Angle sums for triangles

The major result of this section is the following:

Theorem 42. *The sum of the radian measures of the three angles in any triangle is equal to π.*

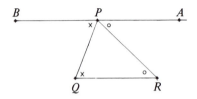

Figure 2.21 Theorem 42. The angle sum of a triangle is π.

Proof: Let PQR be a triangle. Then the unique line through P parallel to $\overleftrightarrow{QR}$ is the union of the rays $\overrightarrow{PA}$ and $\overrightarrow{PB}$, where $A = P + R - Q$ and $B = P + Q - R$. Note that Q is in the interior of $\sphericalangle BPR$ because $Q = B + R - P$. By Theorem 34 the radian measure of $\sphericalangle BPR$ is equal to the sum of the radian measures of $\sphericalangle BPQ$ and $\sphericalangle RPQ$, whereas $\sphericalangle BPR$ and $\sphericalangle APR$ are supplementary. Finally, we apply Theorems 38 and 39 to the alternate angle pairs $\sphericalangle BPQ \simeq \sphericalangle PQR$ and $\sphericalangle APR \simeq \sphericalangle PRQ$ to complete the proof. These constructions are illustrated in Figure 2.21. □

Corollary. *If two angles of a triangle are respectively congruent to two angles of another triangle, then the remaining angles are also congruent.*

Remark: When we study non-Euclidean plane geometry, we will discover that Theorem 42 is one of the few results that is false in non-Euclidean planes. In fact, from an axiomatic approach, this criterion can be used to distinguish Euclidean from non-Euclidean planes.

EXERCISES

1. Find the fixed points and fixed lines of the indicated affine transformations.

   i. $A = \begin{bmatrix} 1 & 2 \\ 2 & 1 \end{bmatrix}$, $b = \begin{bmatrix} 1 \\ -1 \end{bmatrix}$.

   ii. $A = \begin{bmatrix} 1 & 2 \\ 0 & 1 \end{bmatrix}$, $b = \begin{bmatrix} 2 \\ 0 \end{bmatrix}$.

   iii. $A = \begin{bmatrix} 1 & 2 \\ 0 & 1 \end{bmatrix}$, $b = \begin{bmatrix} 2 \\ 1 \end{bmatrix}$.

2. Prove Theorem 4.

3. Let T be a glide reflection with axis m. Show that a line $\ell \neq m$ satisfies $T\ell \| \ell$ if and only if $\ell \| m$ or $\ell \perp m$.

4. Let ℓ be a line of $\mathbf{E}^2$. Let G be the set of isometries T of $\mathbf{E}^2$ satisfying $T\ell = \ell$. Describe the elements of G explicitly and give the group multiplication table.

5. i. Given two intersecting lines, prove that there is a rotation that takes one to the other. Is it unique?
   ii. Given two parallel lines, prove that there is a translation that takes one to the other. Is it unique?

6. Verify that the mapping described following Theorem 5, which associates to each affine transformation a 3 by 3 matrix, is an injective homomorphism of $\mathbf{AF(2)}$ into $\mathbf{GL(3)}$.

7. Prove Theorem 7(ii) by a direct computation using linear algebra.

8. Verify formula (2.4), which shows that every ordinary reflection is an affine reflection.

9. Verify the statements in the remark following Theorem 13.

10. If every nonzero vector in $\mathbf{R}^2$ is an eigenvector of a 2 by 2 matrix A, show that A is a multiple of the identity matrix.

11. Prove Theorem 18.

12. Complete the proof of Theorem 19.

13. Work out the multiplication table for the group of transformations of Theorem 20 and its subgroups.

14. Verify the remark following the definition of radian measure.

15. Prove Theorem 22.

16. Let $P = (1, 2)$, $Q = (0, 0)$, and $R = (2, 1)$. Show that the radian measure of $\angle PQR$ is $\cos^{-1}(4/5)$.

17. Find the bisectors of a straight angle in terms of its vertex and the unit direction vectors of its arms.

18. What is the bisector of a zero angle?

19. If $\overrightarrow{QX}$ is a bisector of an angle $\angle PQR$, prove that $\angle PQX$ and $\angle RQX$ have the same radian measure, namely, half the measure of $\angle PQR$.

20. Prove that the ray $\overrightarrow{QX}$ of Exercise 19 is the only ray having the property described there, unless $\angle PQR$ is a straight angle.

21. Prove that a rotation about P that leaves a ray with origin P fixed must be the identity.

22. Compute the centroid of the following set of points: $\{(1, 4), (2, 4), (1, 0), (2, 0), (9, 7)\}$.

23. Prove Theorem 28. Fill in the details omitted in the text.

24. Prove that there is a reflection interchanging any two lines. Is it unique?

25. Prove that barycentric coordinates are well-defined.

26. With respect to the triangle of reference $P = (-1, 0)$, $Q = (1, 0)$, $R = (0, 1)$, find the barycentric coordinates of the points: $(0, 0)$, $(1, 1)$, $(\sqrt{2}, \sqrt{2})$, $(0, 5)$, $(2, -1)$, $(-\frac{1}{2}, -\frac{1}{3})$.

27. Let a, b, and c be three numbers, not all zero. Show that the set of all points whose barycentric coordinates satisfy $a\lambda + b\mu + c\nu = 0$ is a line.

28. Let P be a point and N a unit vector. Show that $\{X|\langle X - P, N\rangle \geq 0\}$ is a half-plane.

29. Prove Theorem 32.

30. Prove that every point in the interior of an angle lies on a segment joining points of the arms of the angle.

31. Prove that every point in the interior of a triangle lies on a segment joining points on two sides of the triangle.

32. If PQ and $P'Q'$ are collinear segments of equal length, prove that either $\tau_{P'-P}$ or $H_{P'}\tau_{P'-P}$ takes PQ to $P'Q'$.

33. Let P, Q, R, and R' be four distinct points such that $d(P, R) = d(P, R')$ and $d(Q, R) = d(Q, R')$. Prove that $R' = \Omega_\ell R$, where $\ell = \overleftrightarrow{PQ}$.

34. Let T be an affine transformation and ℓ a line. Prove that the points $M = \frac{1}{2}(P + TP)$ (as P ranges through ℓ) are all distinct and collinear, or that they all coincide. Express the locus of M (i.e., the line or

point) in terms of the data determining T and ℓ. This result is called Hjelmslev's theorem, although most treatments assume that T is an isometry.

35. If $[C_1; P_1 \rightarrow Q_1] = [C_2; P_2 \rightarrow Q_2]$, what relationships must hold among the points in question?

36. Show that Theorem 27 and its corollary hold true for affine transformations and affine symmetries.

37. Suppose that an affine transformation has three concurrent fixed lines. Prove that it is a central dilatation or the identity.

38. Extending the notation of Theorem 13, let
$$s_{\lambda,v} = \tau_v s_\lambda, \quad \text{for} \quad \lambda \in \mathbf{R} \text{ and } v \in \mathbf{R}^2.$$

    i. Verify the identity
$$s_{\lambda,u} s_{\mu,v} = s_{\lambda+\mu,w},$$
    where $w = s_{\lambda,u} v$. Thus show that the set of all $s_{\lambda,v}$ is a group.

    ii. Show that $s_{\lambda,v}$ is a shear with horizontal axis if and only if $[v] = [\varepsilon_1]$.

    iii. Show that every shear with horizontal axis may be written in the form $s_{\lambda,v}$.

    iv. Show that $\{s_{\lambda,v} | \langle v, \varepsilon_2 \rangle = 0\}$ is a group (the case $\lambda = 0$ is included here).

Remark: The group defined in Exercise 38(i) is called the *Galilean group* **GAL(2)**. It arises in classical kinematics when describing uniform motion in a straight line. The subset of affine geometry consisting of those facts of Euclidean geometry that continue to make sense when the figure in question is subjected to transformations by the Galilean group is called Galilean geometry and is the subject of an interesting book by I. M. Yaglom [34].

39. Let ∂_1 and ∂_2 be parallel segments. Find a central dilatation taking ∂_1 to ∂_2
    i. by a geometric construction,
    ii. in terms of the end points of the given segments.
    Can there be more than one such dilatation?

40. Prove that any affine transformation that preserves perpendicularity must be a similarity.

41. Prove Pasch's theorem: If a line intersects one side of a triangle, it must also intersect one of the other sides.

42. Let PQR be a triangle. Let F be the foot of the perpendicular from P

to $\overleftrightarrow{QR}$. Prove that F is between Q and R if and only if $\angle PQR$ and $\angle PRQ$ are acute angles.

43. Let P and Q be distinct points. Show that $\overrightarrow{PQ} \cap \overrightarrow{QP} = PQ$.

44. Let T be an affine transformation taking $\triangle PQR$ to $\triangle P'Q'R'$ as in Theorem 36. Prove the following facts:

   i. For any $X \in \mathbf{E}^2$ the barycentric coordinates of TX with reference to $\triangle P'Q'R'$ are the same as those of X with reference to $\triangle PQR$.

  ii. If

$$
\begin{aligned}
P' &= a_{11}P + a_{21}Q + a_{31}R, \\
Q' &= a_{12}P + a_{22}Q + a_{32}R, \\
R' &= a_{13}P + a_{23}Q + a_{33}R,
\end{aligned}
$$

then the barycentric coordinates of $X' = TX$ (with reference to $\triangle PQR$) are related to those of X by the equation

$$
\begin{bmatrix} \lambda' \\ \mu' \\ \nu' \end{bmatrix} = \begin{bmatrix} a_{11} & a_{12} & a_{13} \\ a_{21} & a_{22} & a_{23} \\ a_{31} & a_{32} & a_{33} \end{bmatrix} \begin{bmatrix} \lambda \\ \mu \\ \nu \end{bmatrix}.
$$

45. Prove Heron's theorem: Let ℓ be a line, and let A and B be points not on ℓ. Among all points X on ℓ, the quantity $d(A, X) + d(X, B)$ is minimum when X is on the segment AB or X is on the segment AB', where $B' = \Omega_\ell B$.

Finite groups of isometries of $\mathbf{E}^2$

3

Introduction

So far we have been studying particular figures or transformations with only secondary emphasis on the group structures involved. We shall now turn our attention to the study of groups of transformations and apply our results to get geometrical conclusions. Our main result is to determine precisely the finite groups of isometries of $\mathbf{E}^2$.

Cyclic and dihedral groups

Let m be any positive integer, and let $\alpha = \text{rot}\,(2\pi/m)$. The smallest subgroup of $\mathbf{O(2)}$ containing α is denoted by $\mathbf{C}_m$. We observe that $\mathbf{C}_1 = \{I\}$, but if $m > 1$ then $\mathbf{C}_m$ consists of the distinct elements $I, \alpha, \alpha^2, \ldots, \alpha^{m-2}, \alpha^{m-1}$, because $\alpha^m = I$. Any group isomorphic to $\mathbf{C}_m$ is called a *cyclic* group of order m.

Now let $\beta = \text{ref}\,0$. The smallest subgroup of $\mathbf{O(2)}$ containing both α and β is denoted by $\mathbf{D}_m$.

Theorem 1. *In the group* $\mathbf{D}_m$ *the identity* $\beta\alpha = \alpha^{-1}\beta$ *holds.*

Proof: By Theorem 1.30 we have, for any θ, ϕ,

$$\text{ref }\theta\ \text{rot }\phi = \text{ref}\!\left(\theta - \frac{\phi}{2}\right) = \text{ref}\!\left(\theta + \left(\frac{-\phi}{2}\right)\right) = \text{rot}\,(-\phi)(\text{ref }\theta).$$

Setting $\phi = 2\pi/m$ and $\theta = 0$ yields the desired result. $\qquad\square$

Theorem 2. *The index* $[\mathbf{D}_m : \mathbf{C}_m]$ *of* $\mathbf{C}_m$ *in* $\mathbf{D}_m$ *is equal to* 2.

Proof: The group $\mathbf{C}_m$ consists of m distinct rotations. The coset $\beta\mathbf{C}_m$ consists of m distinct reflections. Because the identity

$$\beta\alpha^j\beta\alpha^k = \alpha^{-j}\beta^2\alpha^k = \alpha^{k-j}$$

holds, the union of the cosets $\beta\mathbf{C}_m$ and $\mathbf{C}_m$ is the group $\mathbf{D}_m$. $\square$

Any group isomorphic to $\mathbf{D}_m$ is called a *dihedral* group.

Remark:
i.   The symmetry group of an angle (other than a straight angle) is isomorphic to $\mathbf{D}_1$. It consists of the identity and a reflection.
ii. The Klein four-group (which we obtained in Theorem 2.28 as the symmetry group of a segment) is isomorphic to $\mathbf{D}_2$. It consists of the identity, a half-turn, and two reflections.
iii. The symmetry group of an equilateral triangle (Theorem 2.37) is isomorphic to $\mathbf{D}_3$.

Conjugate subgroups

Let G be a group. Two subgroups H and K are said to be *conjugate in G* if there exists an element $g \in G$ such that $K = g^{-1}Hg$.

Theorem 3.   *Let g and T be isometries of $\mathbf{E}^2$. Then*
i.   *If T is a reflection, so is $g^{-1}Tg$.*
ii. *If T is a rotation, so is $g^{-1}Tg$.*
iii. *If T is a translation, so is $g^{-1}Tg$.*

Proof:
i.   Let $T = \Omega_\ell$ be a reflection. For any $x \in g^{-1}\ell$,

$$g^{-1}\Omega_\ell gx = g^{-1}gx = x,$$

so that $g^{-1}\ell$ is pointwise fixed. On the other hand, $g^{-1}\Omega_\ell g$ cannot be equal to the identity. Because it has this particular fixed point behavior, $g^{-1}\Omega_\ell g$ must in fact be the reflection in the line $g^{-1}\ell$ (Theorem 1.39).
ii. Suppose that $T = \Omega_\ell\Omega_m$ is a rotation with center P. Then

$$g^{-1}Tg = (g^{-1}\Omega_\ell g)(g^{-1}\Omega_m g),$$

a rotation about $g^{-1}P$.
iii. Suppose that $T = \Omega_\ell\Omega_m$ is a translation. Then $g^{-1}Tg$ is the product of reflections in the parallel lines $g^{-1}\ell$ and $g^{-1}m$. $\square$

Groups of isometries that are conjugate in $\mathscr{I}(\mathbf{E}^2)$ are said to be *geometrically equivalent*. This definition is motivated by observations of the type made in Theorem 3. Conjugate elements perform the "same" isometries with respect to "different" positions. The group of all rotations about a point P is conjugate to the group of all rotations about any other point Q.

An example of two groups that are algebraically equivalent (i.e., isomorphic) but not geometrically equivalent (conjugate) is given by $\mathbf{D}_1$ and $\mathbf{C}_2$. Because both have order 2, they are isomorphic. But $\mathbf{C}_2$ is generated by a half-turn α, and $\mathbf{D}_1$ by a reflection β. Any equation of the form $g^{-1}\alpha g = \beta$ would be impossible because β has a line of fixed points, but α (and hence $g^{-1}\alpha g$) has only one fixed point.

The groups $\mathbf{C}_4$ and $\mathbf{D}_2$ have the same order but are not isomorphic. This can be seen by observing that every element $g \in \mathbf{D}_2$ satisfies the relation $g^2 = I$, which is not satisfied by $\alpha \in \mathbf{C}_4$.

The groups $\mathbf{C}_{2m}$ and $\mathbf{D}_m$ have the same number of elements for each m. If $m > 2$, $\mathbf{C}_{2m}$ is abelian, but $\mathbf{D}_m$ is not. Thus, $\mathbf{D}_m$ and $\mathbf{C}_{2m}$ are never geometrically equivalent and are isomorphic only if $m = 1$.

Theorem 4. *Let $\alpha = \mathrm{rot}\,(2\pi/m)$, $\beta = \mathrm{ref}\,0$, and $\gamma = \mathrm{ref}\,\theta$. Then the group $\langle\{\alpha, \gamma\}\rangle$ generated by α and γ is conjugate to the dihedral group $\mathbf{D}_m = \langle\{\alpha, \beta\}\rangle$.*

Proof: Intuitively speaking, we can realize a reflection in the mirror of γ by first rotating by $-\theta$, then reflecting in the x_1-axis, and finally rotating back by θ. Algebraically, by Theorem 1.30

$$\mathrm{rot}\,\theta\ \mathrm{ref}\,0\ \mathrm{rot}\,(-\theta) = \mathrm{rot}\,\theta\ \mathrm{ref}\,\frac{\theta}{2}$$

$$= \mathrm{ref}\left(\frac{\theta}{2} + \frac{\theta}{2}\right) = \mathrm{ref}\,\theta. \qquad \square$$

It is not surprising that congruent figures have geometrically equivalent symmetry groups.

Theorem 5. *Let $\mathscr{F}$ be a figure, and let g be an isometry of $\mathbf{E}^2$. Then $\mathscr{S}(\mathscr{F})$ and $\mathscr{S}(g\mathscr{F})$ are conjugate in $\mathscr{I}(\mathbf{E}^2)$.*

Proof: Let h be a symmetry of $\mathscr{F}$. Then

$$ghg^{-1}g\mathscr{F} = gh\mathscr{F} = g\mathscr{F},$$

and, hence, ghg^{-1} is a symmetry of $g\mathscr{F}$. Conversely, if h is a symmetry of $g\mathscr{F}$, we have

$$g^{-1}hg\mathscr{F} = g^{-1}g\mathscr{F} = \mathscr{F}.$$

Thus, $g^{-1}\mathscr{S}(g\mathscr{F})g = \mathscr{S}(\mathscr{F})$. $\qquad \square$

Remark: For economy of language we will sometimes substitute "is" for "is geometrically equivalent to" when speaking of groups of isometries. For instance, we say "the symmetry group of the equilateral triangle is $\mathbf{D}_3$," even though this is strictly true only for an equilateral triangle with centroid at the origin and with the x_1-axis as a line of symmetry.

Orbits and stabilizers

Let X be a set, and let G be a group of transformations of X. Let x be a member of X. Then

$$Gx = \{gx | g \in G\}$$

is called the *orbit* of x by G. When G is understood from the context, we may write $\text{Orbit}(x)$ for Gx. The set

$$G_x = \{g \in G | gx = x\}$$

is called the *stabilizer* of x in G. The stabilizer is also sometimes written $\text{Stab}(x)$.

Theorem 6.   *For any $x \in X$ the stabilizer of x in G is a group. There is a natural bijection of the set of cosets determined by $\text{Stab}(x)$ onto $\text{Orbit}(x)$.*

Proof: The fact that G_x is a group is easy. Now consider the mapping $\tau : G \to \text{Orbit}(x)$ defined by $\tau g = gx$. Then τ is clearly surjective. Let $\pi : G \to G/G_x$ be the natural homomorphism. We now set $\bar{\tau}\pi g = \tau g$ and note that $\bar{\tau}$ is well-defined as a map from G/G_x to Gx. To check this, suppose that $\pi g = \pi \bar{g}$ are two representations of an element of G/G_x. Because g and $\bar{g}$ belong to the same coset, they satisfy $\bar{g}^{-1}gx = x$. But then

$$\tau g = gx = (\bar{g}\bar{g}^{-1})gx = \bar{g}(\bar{g}^{-1}g)x = \bar{g}x = \tau\bar{g}.$$

Hence, $\bar{\tau}$ is well-defined. Surjectivity is clear. Furthermore, $\bar{\tau}$ is injective because $\bar{\tau}\pi g = \bar{\tau}\pi\bar{g}$ means that $\tau g = \tau\bar{g}$. In other words,

$$gx = \bar{g}x, \quad \bar{g}^{-1}gx = x, \quad \bar{g}^{-1}g \in G_x, \quad \text{and} \quad \pi g = \pi\bar{g}. \qquad \square$$

Theorem 6 allows us to complete our earlier assertions about symmetry groups of rectilinear figures.

Theorem 7.   *If $\mathscr{F}$ is a rectilinear figure having just one vertex, then $\mathscr{S}(\mathscr{F})$ is a finite group.*

Proof: Let P be the only vertex of $\mathscr{F}$. Then $\hat{\mathscr{F}}$ contains a finite number of lines (say m) through P. Note that $m \geq 2$. Let Q be some point of $\mathscr{F}$ other than P. Then the stabilizer of Q, consisting of isometries leaving P and Q fixed (Theorem 1.39) has at most two elements: the identity and possibly the reflection in $\overleftrightarrow{PQ}$. The orbit of Q by $\mathscr{S}(\mathscr{F})$ consists of points on $\mathscr{F}$ whose distance from P is $d(P, Q)$. Because $\hat{\mathscr{F}}$ has only m lines through P, there can be at most $2m$ such points. Thus

$$\#\mathscr{S}(\mathscr{F}) \leq \#\mathrm{Stab}(Q)\cdot\#\mathrm{Orbit}(Q) \leq 2(2m) = 4m.$$

This estimate is the best possible because a regular polygon with $2m$ vertices determines m lines through its center and has symmetry group of order $4m$. The figure consisting of these m lines alone has the same symmetry group. (Regular polygons are discussed in the next section.) $\square$

Remark: Let $\mathscr{F}$ be a figure consisting of m rays with a common origin P. Then $\mathscr{S}(\mathscr{F})$ has at most $2m$ elements.

Definition.   *Let G be a group of transformations of a set X. If there is a point $x_0 \in X$ that is a fixed point of every transformation in G, we call x_0 a fixed point of G.*

Remark: The orbit of the fixed point x_0 is $\{x_0\}$. The stabilizer of x_0 is the whole group G

Theorem 8.   *Let $\mathscr{F}$ be a rectilinear figure with at least one vertex. Then $\mathscr{S}(\mathscr{F})$ has a fixed point.*

Leonardo's theorem

We now turn to the question, what finite groups can occur as symmetry groups of figures? The answer was known to Leonardo da Vinci ([31], p. 99). Although groups had not been invented in his day, he was aware that the only symmetries of a finite figure were rotations about a certain point and reflections in lines through that point.
   We noted that symmetry groups have the fixed point property. We now show that all finite isometry groups have that property.

Theorem 9.   *Let G be a finite subgroup of $\mathbf{AF(2)}$. Then G has a fixed point.*

Proof: Choose a point $x \in \mathbf{E}^2$. Let $n = \#G$, and $C = (1/n)\Sigma_{g \in G}gx$. That is, C is the centroid of the orbit of x. Any $T \in G$ permutes the elements of the orbit of x. A calculation similar to that of Theorem 2.27 gives us the fact that $TC = C$. $\square$

Corollary. *Every finite subgroup of $\mathscr{I}(\mathbf{E}^2)$ consists of rotations about a certain point and reflections in lines through that point.*

Theorem 10. *Every finite subgroup G of $\mathscr{I}(\mathbf{E}^2)$ is cyclic or dihedral. If C is a fixed point of G, then G is generated by a rotation about C (possibly trivial) and/or a reflection in a line through C.*

Proof: First consider the case $C = 0$. Then G is a subgroup of $\mathbf{O(2)}$. If $G \cap \mathbf{SO(2)} = \{I\}$, then either $G = \{I\}$ or $G = \{I, \beta\}$ (where β is some reflection), because if G contained more than one reflection, it would have to contain a nontrivial rotation as well.

Suppose now that G contains a nontrivial rotation. Let ϕ be the smallest positive number such that rot $\phi \in G$. If rot ψ is another element of G, we may choose an integer ℓ so that

$$\ell\phi \le \psi < (\ell + 1)\phi;$$

that is,

$$0 \le \psi - \ell\phi < \phi.$$

Now $\mathrm{rot}\,(\psi - \ell\phi) = (\mathrm{rot}\,\psi)(\mathrm{rot}\,\phi)^{-\ell}$ is a member of G. Hence, $\psi - \ell\phi = 0$, and we conclude that all rotations in G are powers of rot ϕ. The same calculation with $\psi = 2\pi$ shows that there is a positive integer m such that $m\phi = 2\pi$; that is, $\phi = 2\pi/m$. Thus, we have shown that $G \cap \mathbf{SO(2)} = \mathbf{C}_m$.

Now if G contains a reflection β, we see that the coset $\mathbf{C}_m\beta$ contains m distinct reflections. Every reflection in G lies in $\mathbf{C}_m\beta$ because if γ is such a reflection then $\gamma\beta$ is a rotation in $\mathbf{C}_m$ and $\gamma = (\gamma\beta)\beta \in \mathbf{C}_m\beta$. Thus, every element of G may be written in the form $\alpha^j\beta^k$, where $\alpha = \mathrm{rot}\,(2\pi/m)$, $0 \le j < m$, and $0 \le k < 2$.

The identity $\beta\alpha = \alpha^{-1}\beta$ can be easily verified as in Theorem 1, and thus G is either $\mathbf{C}_m$ or $\mathbf{D}_m$.

Finally, if C is a point other than the origin, G is conjugate to the group $\tau_{-C}G\tau_C$, which does leave the origin fixed, and the first part of the proof applies. $\qquad\square$

Regular polygons

As we have just discovered, only the cyclic and dihedral groups can occur as symmetry groups of figures. We will now describe a family of figures having precisely these symmetry groups. These are based on the familiar notion of regular polygons. Intuitively, we may imagine tracing out a figure by moving a unit distance, then turning through an angle of $2\pi/m$, and repeating this process m times. (This is the "turtle geometry" approach [1].)

We now make the formal definition.

Definition. *Let m be a positive integer greater than 2. Let P and Q be distinct points of* $\mathbf{E}^2$. *For each integer k let*

$$Q_k = \tau_P\left(\mathrm{rot}\ \frac{2\pi k}{m}\right)\tau_{-P}Q,$$

and let q_k *be the segment* Q_kQ_{k+1}. *The union of all* q_k *is called a* regular polygon. *See Figures 3.1–3.2.*

Observing that $Q_{k+m} = Q_k$ and $q_{k+m} = q_k$ for all k, we see that there are m distinct Q_k (called vertices of the polygon) and m distinct q_k (called edges or sides of the polygon). The expression "polygon with m sides" is sometimes abbreviated "m-gon." But a regular polygon also is a rectilinear figure, and the concept of vertex has already been defined in Chapter 2. We must show that the two definitions are consistent. As a first step, we show that the centroid of the Q_k is P.

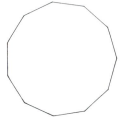

Figure 3.1 A regular 11-gon – symmetry group is dihedral.

Theorem 11. *Let* $\{Q_k\}$ *be the vertices of a regular m-gon. Then*

$$\frac{1}{m} \sum_{k=1}^{m} Q_k = P.$$

Proof: First note that

$$Q_k = P + \alpha^k(Q - P),$$

where $\alpha = \mathrm{rot}\,(2\pi/m)$. Thus,

$$\Sigma Q_k = mP + (\Sigma\alpha^k)(Q - P).$$

Figure 3.2 A regular 7-gon – symmetry group is dihedral.

The proof normally used to derive the formula for the sum of a geometric series applies here to show that the matrix equation

$$\sum_{k=1}^{m} \alpha^k = 0$$

holds. This completes the proof. □

Theorem 12. *Let* $\{Q_k\}$ *be the vertices of a regular m-gon. Let* $\ell_k = \overleftrightarrow{Q_kQ_{k+1}}$ *be a line determined by consecutive vertices. Then the whole polygon lies on the same side of* ℓ_k.

Proof: Let $v = Q_k - P$. Then, we may write

$$Q_{k+1} = P + (\mathrm{rot}\ \theta)v, \quad Q_j = P + (\mathrm{rot}\ \delta)v,$$

where $\theta = 2\pi/m$ and $\delta = (j - k)\theta$. Now

$$Q_{k+1} - Q_k = (\mathrm{rot}\ \theta - I)v = 2\left(\sin \frac{\theta}{2}\right)J\left(\mathrm{rot}\ \frac{\theta}{2}\right)v$$

and

$$Q_j - Q_k = 2\left(\sin\frac{\delta}{2}\right)J\left(\operatorname{rot}\frac{\delta}{2}\right)v.$$

The equation of the line ℓ_k is

$$\langle x - Q_k, N\rangle = 0,$$

where $N = J(Q_{k+1} - Q_k)$. We compute

$$\langle Q_j - Q_k, N\rangle = 4\sin\frac{\theta}{2}\sin\frac{\delta}{2}\left\langle J\left(\operatorname{rot}\frac{\delta}{2}\right)v, -\left(\operatorname{rot}\frac{\theta}{2}\right)v\right\rangle$$

$$= 4\sin\frac{\theta}{2}\sin\frac{\delta}{2}\left\langle\left(\operatorname{rot}\frac{\delta-\theta}{2}\right)v, Jv\right\rangle$$

because $J^2 = -I$ and $J = \operatorname{rot}(\pi/2)$ commutes with the other rotations. But

$$\left(\operatorname{rot}\frac{\delta-\theta}{2}\right)v = \left(\left(\cos\frac{\delta-\theta}{2}\right)I + \left(\sin\frac{\delta-\theta}{2}\right)J\right)v,$$

therefore,

$$\langle Q_j - Q_k, N\rangle = 4\left(\sin\frac{\theta}{2}\right)\left(\sin\frac{\delta}{2}\right)\left(\sin\frac{\delta-\theta}{2}\right)|Jv|^2.$$

At this stage we remark that $\sin(\delta/2)$ and $\sin((\delta - \theta)/2)$ cannot have different signs. At worst, one can be zero, and the other nonzero. We conclude that $\langle Q_j - Q_k, N\rangle \geq 0$ for all j, and that all the Q_j lie on the same side of ℓ_k. Furthermore, equality occurs if and only if $\delta = 0$ (i.e., $Q_j = Q_k$) or $\delta = \theta$ (i.e., $Q_j = Q_{k+1}$). $\quad\square$

Corollary. *The line ℓ does not intersect any segment of the form Q_jQ_{j+1} except for the following cases:*
i. $Q_j = Q_k$, $Q_{j+1} = Q_{k+1}$,
ii. $Q_j = Q_{k+1}$,
iii. $Q_k = Q_{j+1}$.

Corollary. *The vertices of the polygon are precisely the m points $\{Q_j\}$, $j = 1, 2, \ldots, m$.*

Similarity of regular polygons

When computing the symmetry group of a regular polygon, it is legitimate to assume that its center is at the origin and one of its vertices is at $\varepsilon_1 = (1, 0)$. The reason is that if $\mathscr{P}$ and $\mathscr{P}'$ are two regular m-gons, then $\mathscr{S}(\mathscr{P})$ and $\mathscr{S}(\mathscr{P}')$ are conjugate in $\mathscr{I}(\mathbf{E}^2)$. We will now justify this statement.

Theorem 13. *Any two regular m-gons are similar.*

Proof: Let P, Q and P', Q' be, respectively, the center and a vertex of two regular m-gons $\mathscr{P}$ and $\mathscr{P}'$. Then $\tau_{-P}\mathscr{P}$ and $\tau_{-P'}\mathscr{P}'$ are regular m-gons congruent to $\mathscr{P}$ and $\mathscr{P}'$. Let ϕ and ϕ' be chosen so that $(\text{rot } \phi)\tau_{-P}\mathscr{P}$ and $(\text{rot } \phi')\tau_{-P'}\mathscr{P}'$ are regular m-gons centered at the origin with one vertex on the positive x_1-axis (i.e., the ray with origin 0 and direction vector ε_1). Call the new polygons $\mathscr{P}_0$ and $\mathscr{P}'_0$.

Now if $Q = (q, 0)$ and $Q' = (q', 0)$, let S be the central dilatation given by

$$Sx = \frac{q'}{q}x.$$

We claim that $S\mathscr{P}_0 = \mathscr{P}'_0$. First note that if Q_j and Q'_j are the vertices of $\mathscr{P}_0$ and $\mathscr{P}'_0$, respectively, then

$$SQ_j = S(\text{rot } j\theta)Q = (\text{rot } j\theta)\frac{q'}{q}Q$$

$$= (\text{rot } j\theta)Q' = Q'_j,$$

where $\theta = 2\pi/m$. Thus if q_j and q'_j are the edges of $\mathscr{P}_0$ and $\mathscr{P}'_0$, respectively, then

$$Sq_j = q'_j.$$

We conclude that $S\mathscr{P}_0 = \mathscr{P}'_0$. Putting all this together we see that

$$S(\text{rot } \phi)\tau_{-P}\mathscr{P} = (\text{rot } \phi')\tau_{-P'}\mathscr{P}';$$

that is,

$$\tau_{P'}(\text{rot}(-\phi'))S(\text{rot } \phi)\tau_{-P}\mathscr{P} = \mathscr{P}'. \qquad \square$$

Theorem 14. *The symmetry groups of any two regular m-gons are conjugate in $\mathscr{I}(\mathbf{E}^2)$.*

Proof: Let T be the similarity constructed in the previous theorem. Then $\mathscr{S}(\mathscr{P})$ and $\mathscr{S}(\mathscr{P}')$ are conjugate in the group of similarities. This is not enough. However, if $g \in \mathscr{S}(\mathscr{P})$, then

$$g = \tau_P g_0 \tau_{-P},$$

where $g_0 \in \mathbf{O}(2)$. This is because g leaves P fixed. Now

$$\begin{aligned}
TgT^{-1} &= \tau_{P'}(\text{rot}(\phi - \phi'))S\tau_{-P}\tau_P g_0 \tau_{-P}\tau_P S^{-1}(\text{rot}(\phi' - \phi))\tau_{-P'} \\
&= \tau_{P'}(\text{rot}(\phi - \phi'))Sg_0 S^{-1}(\text{rot}(\phi' - \phi))\tau_{-P'}.
\end{aligned}$$

We observe that S commutes with every member of $\mathbf{O}(2)$. Hence,

$$TgT^{-1} = \tau_{P'}(\text{rot}(\phi - \phi'))\tau_{-P}g\tau_P(\text{rot}(\phi' - \phi))\tau_{-P'} = \tilde{T}g\tilde{T}^{-1},$$

where $\bar{T}$ is an isometry. This shows that $\mathscr{S}(\mathscr{P})$ and $\mathscr{S}(\mathscr{P}')$ are in fact conjugate in $\mathscr{I}(\mathbf{E}^2)$.

Symmetry of regular polygons

Theorem 15. *Let $\mathscr{P}$ be a regular m-gon. Then $\mathscr{S}(\mathscr{P}) = \mathbf{D}_m$.*

Proof: By Theorems 12 and 13 we may assume that the vertices of $\mathscr{P}$ are of the form

$$\left(\mathrm{rot}\frac{2\pi k}{m}\right)\varepsilon_1, \quad k = 0, 1, 2, \ldots, m - 1.$$

Clearly, $\mathrm{rot}(2\pi/m)$ permutes the vertices. In fact,

$$\left(\mathrm{rot}\,\frac{2\pi}{m}\right)Q_j = Q_{j+1}.$$

Secondly, ref 0 permutes the vertices. Specifically,

$$(\mathrm{ref}\ 0)Q_j = Q_{m-j}.$$

Adjacent vertices remain adjacent under both of these transformations. In particular,

$$\left(\mathrm{rot}\,\frac{2\pi}{m}\right)q_j = q_{j+1}, \quad (\mathrm{ref}\ 0)\ q_j = q_{m-j-1}.$$

Thus, the edges are permuted by $\mathrm{rot}\,(2\pi/m)$ and ref 0. Because each of these transformations is in $\mathscr{S}(\mathscr{P})$, we must conclude that $\mathbf{D}_m \subset \mathscr{S}(\mathscr{P})$. But Leonardo's theorem shows that $\mathscr{S}(\mathscr{P})$ is cyclic or dihedral because the centroid must be left fixed. If $\mathscr{S}(\mathscr{P})$ contains a rotation other than those in $\mathbf{C}_m$ (call it rot θ), then rot θ must permute the vertices. In particular, (rot θ)Q must be a vertex and so must be equal to $\mathrm{rot}(2\pi j/m)\varepsilon_1$ for some j. Thus, rot $\theta \in \mathbf{C}_m$, and the proof is complete. $\quad\square$

Leonardo's theorem together with Theorem 7 shows that the only groups that can occur as symmetry groups of rectilinear figures (having at least one vertex) are $\mathbf{D}_m$ and $\mathbf{C}_m$.

The work on regular polygons shows that every dihedral group can be obtained as the symmetry group of some figure.

We now ask whether every cyclic group $\mathbf{C}_m$ can occur as the symmetry group of a figure. The answer can be seen as follows. A regular m-gon has $2m$ symmetries, m rotations, and m reflections. We change the figure in such a way as to destroy the bilateral symmetry while retaining the rotational symmetry. This can be done by attaching a tail to one end of each of the edges.

Let Q_j be defined as before, but let q_j be the ray with origin Q_j and direction vector $Q_{j+1} - Q_j$. One can check that no new vertices are introduced by this procedure. See Figures 3.3 and 3.4. However,

$$\left(\operatorname{rot}\frac{2\pi}{m}\right)q_j = q_{j+1} \quad \text{and} \quad \left(\operatorname{rot}\frac{2\pi}{m}\right)Q_j = Q_{j+1}.$$

Thus $\mathbf{C}_m \subset \mathscr{S}(\mathscr{F})$. As before, these are the only rotations in $\mathscr{S}(\mathscr{F})$.

If any reflection ref ϕ were a symmetry of $\mathscr{F}$, it would have to permute the vertices of $\mathscr{F}$ as well as leaving the centroid (the origin in this case) fixed. At most, two vertices could be fixed by ref ϕ. Thus, for some $j \neq k$, we would have

$$(\operatorname{ref}\phi)Q_j = Q_k.$$

Then

$$\begin{aligned}(\operatorname{ref}\phi)Q_{j+1} &= \operatorname{ref}\phi(\operatorname{rot}\theta)Q_j = \operatorname{rot}(-\theta)(\operatorname{ref}\phi)Q_j \\ &= (\operatorname{rot}(-\theta))Q_k = Q_{k-1}.\end{aligned}$$

Thus, ref ϕ sends the ray $\overrightarrow{Q_j Q_{j+1}}$ to the ray $\overrightarrow{Q_k Q_{k-1}}$, which is not in the figure $\mathscr{P}$. We conclude that $\mathscr{S}(\mathscr{P})$ contains no reflections, and $\mathscr{S}(\mathscr{P}) = \mathbf{C}_m$. We can now state

Theorem 16. *Every finite cyclic or dihedral group is the symmetry group of a rectilinear figure.*

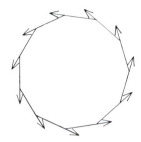

Figure 3.3 A modified regular 11-gon – symmetry group is cyclic.

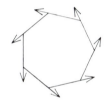

Figure 3.4 A modified regular 7-gon – symmetry group is cyclic.

Figures with no vertices

A complete rectilinear figure with no vertices must consist of a finite number of parallel lines. Let $\mathscr{F}$ be such a figure, and let $[v]$ be the direction of the lines.

Because τ_v is a symmetry of $\mathscr{F}$, we have an example of a rectilinear figure with an infinite symmetry group.

Theorem 17. *Let $\mathscr{F}$ be a figure consisting of a finite number of parallel lines with direction $[v]$. Let $\mathscr{T}$ be the set of translations in $\mathscr{S}(\mathscr{F})$. Then*

$$\mathscr{T} = \{\tau_w | w \in [v]\}.$$

Proof: $\mathscr{F}$ is the union of lines of the form $P + [v]$. Now

$$\tau_w(P + [v]) = P + [v] + w = P + [v].$$

On the other hand, if $\tau_w \in \mathscr{T}$, then $\tau_w P = P + w$ must be in $P + [v]$. Thus $w \in [v]$. $\qquad\qquad\square$

81

Theorem 18.   *Let $\mathscr{F}$ be the figure of Theorem 17. Then $\mathscr{S}(\mathscr{F}) \cap \mathbf{O(2)}$ has at most four elements.*

Proof: Let A be a member of the group in question. Because A permutes the lines of $\mathscr{F}$, we must have $Av = \pm v$. Thus, A is a symmetry of the segment joining v and $-v$. By Theorem 2.28 there are at most four possibilities: two reflections, a half-turn, and the identity. In our case the identity and the reflection that interchanges v and $-v$ necessarily belong to $\mathscr{S}(\mathscr{F})$. The other two transformations will belong only if the lines are placed in a certain way.

Theorem 19.   *Let $\mathscr{F}$ be a complete rectilinear figure with no vertices, and let $\mathscr{T}$ be the set of translations in $\mathscr{S}(\mathscr{F})$. Then*
 i.   *$\mathscr{T}$ is a normal subgroup of $\mathscr{S}(\mathscr{F})$,*
 ii.   *$\mathscr{S}(\mathscr{F})/\mathscr{T}$ has at most four elements.*

Proof: The homomorphism that sends each symmetry onto its linear part has $\mathscr{T}$ as its kernel and $\mathscr{S}(\mathscr{F}) \cap \mathbf{O(2)}$ as its range. Thus, $\mathscr{T}$ is normal, and the quotient group is isomorphic to $\mathscr{S}(\mathscr{F}) \cap \mathbf{O(2)}$.

EXERCISES

1.   A *parallelogram* is a rectilinear figure consisting of four segments (sides) AB, BC, CD, and DA, where $AB\|CD$ and $BC\|DA$. Prove that there is an affine transformation relating any two parallelograms.

2.   Find the affine symmetry group of the parallelogram.

3.   A *rhombus* is a parallelogram in which all four sides have equal lengths. A *rectangle* is a parallelogram in which adjacent sides are perpendicular. A parallelogram that is both a rhombus and a rectangle is called a *square*. Find the symmetry groups of all types of parallelograms.

4.   Let ℓ be a line. Show that $\mathrm{TRANS}(\ell) \cup \{H_P | P \in \ell\}$ is a group, and write down a multiplication table for it. Show that $\mathrm{TRANS}(\ell)$ is a normal subgroup, and describe the quotient group.

5.   Let P and Q be distinct points. Describe the group G generated by $\{H_P, H_Q\}$. Show that the set of translations in G is a normal subgroup and describe the quotient group.

6.   Let $\mathscr{F}$ be the union of three segments AB, BC, and CD. Given that $d(A, B) = d(C, D)$, $AB \perp BC$, and $BC \perp CD$, what can you say about $\mathscr{S}(\mathscr{F})$?

7.   Let $\mathscr{F}$ be a complete rectilinear figure having two or more vertices, all of which are collinear. Show that $\mathscr{S}(\mathscr{F})$ is a finite group having one,

two, or four elements. Describe the configuration in each of these cases.

8. Verify the equation $\sum_{k=1}^{m} \alpha^k = 0$ in Theorem 11.

4

Geometry on the sphere

Introduction

We now turn to a study of spherical geometry. Although the analytic geometry of the sphere is best formulated by considering it as a subset of three-dimensional space, our intuitive motivation must be *intrinsic*. In other words, our geometrical statements must be concerned with the sphere itself, not the points of space that lie inside or outside it. Our point of view is that of a small bug crawling on the two-dimensional surface of the sphere. Concepts of point, line, distance, angle, and reflection will be chosen to coincide with the bug's experience. (See [1], Chapter 5; [30], Chapter 2.)

Preliminaries from $\mathbf{E}^3$

Of course, there is a three-dimensional Euclidean geometry analogous to the geometry of $\mathbf{E}^2$, which is worthy of study in itself. In this book, however, we are restricting our attention to two-dimensional geometries. It is convenient for computational purposes to regard some of these geometries as subsets of $\mathbf{E}^3$, and thus a few facts about the geometry of $\mathbf{E}^3$ will be developed. In a manner quite similar to that used in Chapter 1, we introduce the coordinate three-space $\mathbf{R}^3$ (also a vector space), an inner product, and the concept of length of a vector. In particular, if $x = (x_1, x_2, x_3)$ and $y = (y_1, y_2, y_3)$, then

$$x + y = (x_1 + y_1, x_2 + y_2, x_3 + y_3),$$

$$cx = (cx_1, cx_2, cx_3),$$

$$\langle x, y \rangle = x_1 y_1 + x_2 y_2 + x_3 y_3,$$

$$|x| = \sqrt{\langle x, x \rangle}.$$

Theorems 1–8 of Chapter 1 apply equally well in this setting. The reader can easily check this by using the same proofs or trivial modifications of them.

The theorem of Pythagoras (Theorem 12, Chapter 1) is equally valid in $\mathbf{E}^3$ with the same proof. The definition of $v^{\perp}$ is peculiar to $\mathbf{E}^2$, however. Instead, we have the cross product, which is treated in the next section.

The cross product

The problem of finding a vector perpendicular to two given vectors is solved as follows:

Definition. *Let u and v be vectors in $\mathbf{R}^3$. Then $u \times v$ is the unique vector z such that, for all $x \in \mathbf{R}^3$,*

$$\langle z, x \rangle = \det(x, u, v).$$

Theorem 1.
  i. $u \times v$ is well-defined.
  ii. $\langle u \times v, u \rangle = \langle u \times v, v \rangle = 0$. *(See Figure 4.1.)*
  iii. $u \times v = -v \times u$.
  iv. $\langle u \times v, w \rangle = \langle u, v \times w \rangle$.
  v. $(u \times v) \times w = \langle u, w \rangle v - \langle v, w \rangle u$.

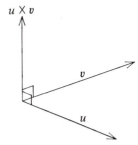

Figure 4.1   The cross product $u \times v$.

Proof: We first recall a result from linear algebra; namely, that every linear function from $\mathbf{R}^3$ to $\mathbf{R}$ can be expressed in the form

$$x \rightarrow \langle x, z \rangle$$

for some fixed vector z (Theorem 8D). As we know, the function

$$x \rightarrow \det(x, u, v)$$

is linear for each fixed choice of u and v. This proves (i). Identities (ii)–(iv) can be easily deduced from the properties of determinants. On the other hand, (v) (often called the vector triple product formula) is rather complicated, but detailed computation can be avoided by exploiting the linearity. First, observe that

$$\varepsilon_1 \times \varepsilon_2 = \varepsilon_3, \quad \varepsilon_2 \times \varepsilon_3 = \varepsilon_1, \quad \text{and} \quad \varepsilon_3 \times \varepsilon_1 = \varepsilon_2.$$

Thus,

$$(\varepsilon_1 \times \varepsilon_2) \times \varepsilon_3 = 0 = \langle \varepsilon_1, \varepsilon_3 \rangle \varepsilon_2 - \langle \varepsilon_2, \varepsilon_3 \rangle \varepsilon_1,$$

$$(\varepsilon_2 \times \varepsilon_3) \times \varepsilon_3 = -\varepsilon_2 = \langle \varepsilon_2, \varepsilon_3 \rangle \varepsilon_3 - \langle \varepsilon_3, \varepsilon_3 \rangle \varepsilon_2,$$

$$(\varepsilon_3 \times \varepsilon_1) \times \varepsilon_3 = \varepsilon_1 = \langle \varepsilon_3, \varepsilon_3 \rangle \varepsilon_1 - \langle \varepsilon_1, \varepsilon_3 \rangle \varepsilon_3.$$

85

By linearity we have

$$(u \times v) \times \varepsilon_3 = \langle u, \varepsilon_3 \rangle v - \langle v, \varepsilon_3 \rangle u.$$

By symmetry the analogous identity is true when ε_3 is replaced by ε_1 or ε_2. Finally, by linearity

$$(u \times v) \times w = \langle u, w \rangle v - \langle v, w \rangle u. \qquad \square$$

Corollary.

i. $u \times v = 0$ *if and only if u and v are proportional.*

ii. *If* $u \times v \neq 0$, *then* $\{u, v, u \times v\}$ *is a basis for* $\mathbf{R}^3$.

iii. $\langle u \times v, w \times z \rangle = \langle u, w \rangle \langle v, z \rangle - \langle v, w \rangle \langle u, z \rangle.$

iv. $|u \times v|^2 = |u|^2|v|^2 - \langle u, v \rangle^2.$

This last statement is known as the Lagrange identity. Note that it yields another proof of the Cauchy–Schwarz inequality.

Proof: (i), (iii), and (iv) can be easily deduced from the results of the theorem. (See Exercise 1.) For (ii) we will show that the set of vectors in question is linearly independent. Then general results from linear algebra (Appendix D) can be applied.

Now if there exist numbers λ, μ, ν with

$$\lambda u + \mu v + \nu(u \times v) = 0,$$

we can take inner product with $u \times v$ to obtain

$$\nu|u \times v|^2 = 0,$$

and, hence, $\nu = 0$. Further, taking cross products with v and u, respectively, yields

$$\lambda(u \times v) = 0, \quad \mu(v \times u) = 0,$$

so that $\lambda = \mu = 0$. $\qquad \square$

Orthonormal bases

A triple $\{u, v, w\}$ of mutually orthogonal unit vectors is called an *orthonormal triple*.

Theorem 2.   *If* $\{u, v, w\}$ *is an orthonormal triple, then for all* $x \in \mathbf{R}^3$,

$$x = \langle x, u \rangle u + \langle x, v \rangle v + \langle x, w \rangle w.$$

The proof of this is similar to that of Theorem 9 in Chapter 1. In light of this result we usually refer to such a triple as an orthonormal basis.

Theorem 3. *If u is any unit vector, there exist vectors v and w so that {u, v, w} is an orthonormal basis.*

Proof: Let ξ be any unit vector other than $\pm u$. Then let v be $u \times \xi$ divided by its length, and $w = u \times v$. Noting that $|u \times v|^2 = |u|^2|v|^2 - \langle u, v \rangle^2 = 1$, we see that $\{u, v, w\}$ is orthonormal. □

Planes

A *plane* is a set Π of points of $\mathbf{E}^3$ with the following properties:

i. Π is not contained in any line.
ii. The line joining any two points of Π lies in Π.
iii. Not every point of $\mathbf{E}^3$ is in Π.

Theorem 4.
i.   *If v and w are not proportional, and P is any point, then $P + [v, w]$ is a plane. We speak of the plane through P spanned by $\{v, w\}$.*
ii. *If P, Q, and R are noncollinear points, there is a unique plane Π containing them. In this case we speak of the plane PQR.*
iii. *If N is a unit vector and P is a point, then $\{X | \langle X - P, N \rangle = 0\}$ is a plane. We speak of the plane through P with unit normal N. See Figure 4.2.*

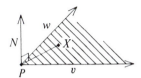

Figure 4.2 X lies on the plane through P with unit normal N.

Notation: $[v, w] = \{tv + sw | t, s \in \mathbf{R}\}$ is called the *span* of $\{v, w\}$.

Proof:
i.   Suppose that $\alpha = P + [v, w]$ is a set as described in (i). We show that α is a plane. First of all, let $Q = P + v$ and $R = P + w$. Then, because $Q - P$ and $R - P$ are not proportional, the points P, Q, and R are not collinear and α is not contained in any line. Secondly, if $X = P + v \times w$, we see that $X \notin \alpha$ because $\{v, w, v \times w\}$ is a linearly independent set. Thus, not every point of $\mathbf{E}^3$ is in α. Thirdly, let

$$X = P + x_1 v + x_2 w, \quad Y = P + y_1 v + y_2 w$$

be points of α, and let t be any real number. Then

$$(1 - t)X + tY = (1 - t)P + tP + ((1 - t)x_1 + ty_1)v$$
$$+ ((1 - t)x_2 + ty_2)w = P + ((1 - t)x_1 + ty_1)v + ((1 - t)x_2 + ty_2)w.$$

This exhibits a typical point of $\overleftrightarrow{XY}$ as a member of α and concludes the proof that α is a plane.
iii. Let P, Q, and R be noncollinear points. Let $v = Q - P$ and $w = R - P$. Then, by (i), $P + [v, w]$ is a plane containing P, Q, and R. We

87

now show that this plane is unique. Let $\tilde{\Pi}$ be any plane containing P, Q, and R. Then

$$
\begin{aligned}
P + \lambda v + \mu w &= P + \lambda(Q - P) + \mu(R - P) \\
&= (1 - \lambda - \mu)P + \lambda Q + \mu R \\
&= (1 - \lambda - \mu)P + (\lambda + \mu)\left(\frac{\lambda}{\lambda + \mu}Q + \frac{\mu}{\lambda + \mu}R\right).
\end{aligned}
$$

This exhibits a typical point of $P + [v, w]$ as a point on the line $\overleftrightarrow{PX}$, where

$$
X = \frac{\lambda}{\lambda + \mu}Q + \frac{\mu}{\lambda + \mu}R
$$

is a point on $\overleftrightarrow{QR}$. Thus, any plane containing P, Q, and R must contain $P + [v, w]$. But now if $\tilde{\Pi}$ contains a point S not in $P + [v, w]$, then $\{S - P, v, w\}$ is a linearly independent set. If Z is any point of $\mathbf{E}^3$, Then

$$
Z - P = \lambda v + \mu w + v(S - P) \quad \text{for some numbers } \lambda, \mu, \text{ and } v.
$$

One can now check that

$$
Z = vS + (1 - v)\left(P + \frac{\lambda}{1 - v}v + \frac{\mu}{1 - v}w\right),
$$

which shows that every point of $\mathbf{E}^3$ is on a line joining S to a point of $\tilde{\Pi}$. This is impossible because $\tilde{\Pi}$ does not contain all of $\mathbf{E}^3$. We conclude that $\tilde{\Pi} = P + [v, w]$ and that the plane containing P, Q, and R is unique.

ii. Finally, we relate characterizations (i) and (ii) of planes. Suppose the unit normal N is given. Then (by Theorem 3) we may choose v and w so that $\{N, v, w\}$ is orthonormal. For any X in $\mathbf{E}^3$ we may write, by Theorem 2,

$$
X - P = \langle X - P, N \rangle N + \langle X - P, v \rangle v + \langle X - P, w \rangle w,
$$

which shows that $X - P$ lies in $[v, w]$ if and only if $\langle X - P, N \rangle = 0$. Thus, $\{X | \langle X - P, N \rangle = 0\}$ is a plane. $\qquad\square$

Incidence geometry of the sphere

The sphere $\mathbf{S}^2$ on whose geometry we will be concentrating is determined by the familiar condition

$$
\mathbf{S}^2 = \{x \in \mathbf{E}^3 | \ |x| = 1\}.
$$

If one begins at a point of $\mathbf{S}^2$ and travels straight ahead on the surface, one will trace out a great circle. Viewed as a set in $\mathbf{E}^3$, this is the intersection of

S^2 with a plane through the origin. However, from the point of view of our bug on S^2 it is more appropriate to call this path a line. This motivates the following definition.

Definition. *Let ξ be a unit vector. Then*

$$\ell = \{x \in S^2 | \langle \xi, x \rangle = 0\}$$

is called the line with pole ξ. *We also call ℓ the* polar line *of ξ.*

Remark: Spherical geometry is non-Euclidean. This means that whenever we represent a figure by a diagram, distortions are inevitable. Diagrams that faithfully represent one aspect (e.g., straightness of lines) will distort some other aspect (e.g., lengths and angles). You are cautioned against basing arguments on a diagram, but you are encouraged to use them to suggest facts that can then be verified rigorously. Often it is desirable to have more than one diagram of the same situation, each providing insight, yet containing some misleading information. Figures 4.3 and 4.4 show two ways of thinking about a point and its polar line.

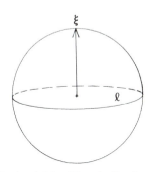

Figure 4.3 A point ξ and its polar line ℓ, first view.

Two points P and Q of S^2 are said to be *antipodal* if $P = -Q$. Lines of S^2 cannot be parallel, and two lines intersect not in just one point but in a pair of antipodal points. We assert the following facts that you may verify as exercises (Exercise 5).

Theorem 5.
  i. *If ξ is a pole of ℓ, so is its antipode $-\xi$.*
 ii. *If P lies on ℓ, so does its antipode $-P$.*

However, once these facts are noticed, there are no further anomalies, and we get the following analogues of the Euclidean results.

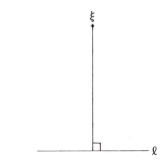

Figure 4.4 A point ξ and its polar line ℓ, second view.

Theorem 6. *Let P and Q be distinct points of S^2 that are not antipodal. Then there is a unique line containing P and Q, which we denote by $\overleftrightarrow{PQ}$.*

Proof: In order to determine a candidate for $\overleftrightarrow{PQ}$, we need a pole ξ. This must be a unit vector orthogonal to both P and Q. Because P and Q are not antipodal, we may choose ξ equal to $(P \times Q)/|P \times Q|$. Clearly, the line with pole ξ passes through P and Q.

We now consider uniqueness. If η is a pole of any line through P and Q, we must have

$$\langle \eta, P \rangle = \langle \eta, Q \rangle = 0.$$

Thus, by the triple product formula, in Theorem 1,

89

$$\eta \times (P \times Q) = 0,$$

and, hence, η is a multiple of the nonzero vector $P \times Q$. Because $|\eta| = 1$, we must have $\eta = \pm\xi$. Thus, $\overleftrightarrow{PQ}$ is uniquely determined. □

Theorem 7. *Let ℓ and m be distinct lines of $\mathbf{S}^2$. Then ℓ and m have exactly two points of intersection, and these points are antipodal. (See Figures 4.5 and 4.6.)*

Proof: Suppose ξ and η are poles of ℓ and m, respectively. Because ℓ and m are distinct, $\xi \neq \pm\eta$, and, hence, $\xi \times \eta \neq 0$. But clearly, both points $\pm(\xi \times \eta)/|\xi \times \eta|$ lie in the intersection. Any third point, however, could lie on at most one of ℓ and m by the uniqueness part of the previous theorem. □

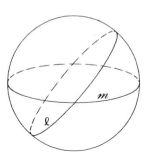

Figure 4.5 Two intersecting lines ℓ and m, first view.

Corollary. *No two lines of $\mathbf{S}^2$ can be parallel.*

Remark: Even lines that have a common perpendicular will intersect. See Figures 4.7 and 4.8 for two views of this situation.

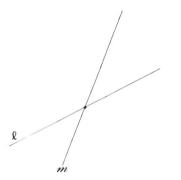

Figure 4.6 Two intersecting lines ℓ and m, second view.

Distance and the triangle inequality

The distance between two points P and Q of $\mathbf{S}^2$ is defined by the equation

$$d(P, Q) = \cos^{-1}\langle P, Q \rangle.$$

This definition reflects the idea that the measure of the angle subtended at the center of the sphere by the arc PQ should be numerically equal to the length of the arc. See Figures 4.9 and 4.10. The following theorem should be compared with Theorem 5 of Chapter 1.

Theorem 8. *If P, Q, and R are points of $\mathbf{S}^2$, then*
  i. $d(P, Q) \geq 0$.
  ii. $d(P, Q) = 0$ *if and only if* $P = Q$.
  iii. $d(P, Q) = d(Q, P)$.
  iv. $d(P, Q) + d(Q, R) \geq d(P, R)$ *(the triangle inequality)*.

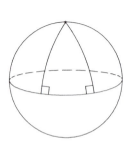

Figure 4.7 Even lines with a common perpendicular are not parallel, first view.

Proof: Properties (i)–(iii) follow from the Cauchy–Schwarz inequality and the properties of the $\cos^{-1}$ function. (See Appendix F.) The details are left to the reader as exercises. We concentrate our attention on the triangle inequality.

Let $r = d(P, Q)$, $p = d(Q, R)$, and $q = d(P, R)$. By the Cauchy–Schwarz inequality we have

$$\langle P \times R, Q \times R \rangle^2 \leqslant |P \times R|^2 \, |Q \times R|^2.$$

Applying Theorem 1, we get that the left side reduces to

$$(\langle P, Q \rangle \langle R, R \rangle - \langle P, R \rangle \langle R, Q \rangle)^2 = (\cos r - \cos q \cos p)^2,$$

and the right side is

$$(1 - \langle P, R \rangle^2)(1 - \langle Q, R \rangle^2) = (1 - \cos^2 q)(1 - \cos^2 p)$$
$$= \sin^2 q \sin^2 p.$$

Thus,

$$\cos r - \cos q \cos p \leqslant \sin q \sin p,$$

and, hence,

$$\cos r \leqslant \cos(q - p).$$

Because the cosine function is decreasing on $[0, \pi]$, we have $r \geqslant q - p$, and, hence, $r + p \geqslant q$, provided that $0 \leqslant q - p \leqslant \pi$. But if $q - p < 0$, then $q < p \leqslant r + p$ in any case. Furthermore, $q - p > \pi$ is impossible.
   We conclude therefore that

$$d(P, Q) + d(Q, R) \geqslant d(P, R),$$

as required. $\qquad \qquad \square$

Corollary. *If equality holds in (iv), then P, Q, and R are collinear.*

Proof: In the proof of (iv), $r = q - p$ implies that the Cauchy–Schwarz inequality is an equality. Thus, $P \times R$ and $Q \times R$ are proportional. Assuming that $P \times R \neq 0$ (otherwise P, Q, and R are automatically collinear), we see that the pole of the line $\overleftrightarrow{PR}$ is proportional to $P \times R$ and, hence, to $Q \times R$. This shows that Q lies on $\overleftrightarrow{PR}$. $\qquad \square$

Remark: In the case of $\mathbf{E}^2$ we get the further conclusion that Q is between P and R (Theorem 1.7). In spherical geometry we shall see that a similar result holds if we make the right definitions. (See Theorem 35 in this chapter.)

Parametric representation of lines

Just as in $\mathbf{E}^2$, it is often convenient to describe lines in parametric form. Suppose that ℓ is a line with pole ξ. Let P and Q be chosen so that $\{\xi, P, Q\}$ is orthonormal. Then set

$$\alpha(t) = (\cos t)P + (\sin t)Q.$$

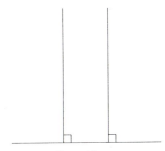

Figure 4.8 Even lines with a common perpendicular are not parallel, second view.

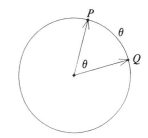

Figure 4.9 Distance in spherical geometry, first view.

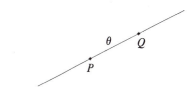

Figure 4.10 Distance in spherical geometry, second view.

91

Theorem 9.

  i. $\ell = \{\alpha(t)|t \in \mathbf{R}\}$.

  ii. *Each point of ℓ occurs exactly once as a value of $\alpha(t)$ while t ranges through the interval $[0, 2\pi)$.*

  iii. $d(\alpha(t_1), \alpha(t_2)) = |t_1 - t_2|$ *if* $0 \leqslant |t_1 - t_2| \leqslant \pi$.

This essentially says that α is a unit-speed parametrization of the line ℓ. See Figures 4.11 and 4.12.

The function α is said to be a standard parametrization of ℓ. Each line has many standard parametrizations. P may be any point on ℓ, and for a given P there are two choices of Q.

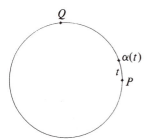

Figure 4.11   Parametrization of a line, first view.

Perpendicular lines

Definition.   *Two lines are* perpendicular *if their poles are orthogonal.*

We recall that in the Euclidean plane a pencil of parallel lines could be regarded as a pencil of lines with a common perpendicular. In the spherical case we have the following situation. The proofs are left to the reader as exercises.

Theorem 10.   *Let ℓ and m be distinct lines of $\mathbf{S}^2$. Then there is a unique line n such that $\ell \perp n$ and $m \perp n$. The intersection points of ℓ and m are the poles of n.*

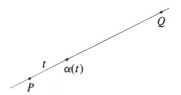

Figure 4.12   Parametrization of a line, second view.

Theorem 11.   *Let ℓ be a line of $\mathbf{S}^2$, and let P be a point. If P is not a pole of ℓ, there is a unique line m through P perpendicular to ℓ.*

Remark:

  i. If P is a pole of ℓ, every line through P will be perpendicular to ℓ.

  ii. Theorem 11 shows that as in $\mathbf{E}^2$, we can drop a perpendicular from a point P not on ℓ to the line ℓ. In $\mathbf{E}^2$ the foot of the perpendicular is the point of ℓ closest to P. In spherical geometry the perpendicular line m intersects ℓ twice. We shall see later in this chapter that the points of intersection are the points of ℓ closest to and farthest from P. We must postpone this discussion, however, until we have more machinery for dealing with segments, angles, and triangles.

Motions of $\mathbf{S}^2$

Definition.   *For any line ℓ the* reflection *in ℓ is the mapping Ω_ℓ given by*

$$\Omega_\ell X = X - 2\langle X, \xi\rangle\xi,$$

It is not obvious that $\Omega_\ell X$ will actually be a point of $\mathbf{S}^2$. To take care of this and other difficulties, we investigate the properties of the transformation of $\mathbf{R}^3$ defined by the formula for Ω_ℓ.

Theorem 12.   *Let $\langle \xi, \xi \rangle = 1$ and define $T: \mathbf{R}^3 \to \mathbf{R}^3$ by*

$$Tx = x - 2\langle x, \xi \rangle \xi.$$

Then

i.   *T is linear.*
ii.   *$\langle Tx, Ty \rangle = \langle x, y \rangle$ for all $x, y \in \mathbf{R}^3$.*

Proof: The linearity of T is an easy verification (Exercise 11); just use the linearity property of the inner product. To see (ii), consider

$$\begin{aligned}
\langle Tx, Ty \rangle &= \langle x - 2\langle x, \xi \rangle \xi, y - 2\langle y, \xi \rangle \xi \rangle \\
&= \langle x, y \rangle - 2\langle x, \xi \rangle \langle \xi, y \rangle - 2\langle x, \xi \rangle \langle y, \xi \rangle + 4\langle x, \xi \rangle \langle y, \xi \rangle \langle \xi, \xi \rangle \\
&= \langle x, y \rangle. \qquad\qquad\qquad\qquad\qquad\qquad\qquad\qquad\qquad\qquad \square
\end{aligned}$$

Remark:

i.   A mapping $T: \mathbf{R}^3 \to \mathbf{R}^3$ satisfying (i) and (ii) in Theorem 12 is said to be orthogonal. We study such mappings in general in a later section.
ii.   Property (ii) says that if $|x| = 1$, then $|Tx| = 1$. Thus, $\Omega_\ell X$ is on $\mathbf{S}^2$ whenever X is a point of $\mathbf{S}^2$. Other consequences of property (ii) and the algebra developed in Chapter 1 yield the following basic properties of reflections.

Theorem 13.

i.   *$d(\Omega_\ell X, \Omega_\ell Y) = d(X, Y)$ for all points X and Y in $\mathbf{S}^2$.*
ii.   *$\Omega_\ell \Omega_\ell X = X$ for all points X in $\mathbf{S}^2$.*
iii.   *$\Omega_\ell: \mathbf{S}^2 \to \mathbf{S}^2$ is a bijection.*

Theorem 14.   *$\Omega_\ell X = X$ if and only if $X \in \ell$.*

We now investigate the product of two reflections. Let ℓ and m be distinct lines with respective poles ξ and η. Let P be one of the points of intersection of ℓ and m. Choose an orthonormal basis $\{e_1, e_2, e_3\}$ with $e_3 = P$. Then ξ and η are unit vectors in the span of $\{e_1, e_2\}$. As in Chapter 1, we may choose numbers θ and ϕ so that

$$\xi = (-\sin \theta)e_1 + (\cos \theta)e_2, \quad \eta = (-\sin \phi)e_1 + (\cos \phi)e_2.$$

A routine calculation [essentially that of Chapter 1, (1.11)] gives

$$\Omega_\ell e_1 = (\cos 2\theta)e_1 + (\sin 2\theta)e_2,$$

$$\Omega_\ell e_2 = (\sin 2\theta)e_1 - (\cos 2\theta)e_2,$$
$$\Omega_\ell e_3 = e_3.$$

Thus, in terms of the basis $\{e_1, e_2, e_3\}$, Ω_ℓ has the matrix

$$\begin{bmatrix} \cos 2\theta & \sin 2\theta & 0 \\ \sin 2\theta & -\cos 2\theta & 0 \\ 0 & 0 & 1 \end{bmatrix},$$

which we may abbreviate as

$$\begin{bmatrix} \text{ref } \theta & 0 \\ 0 & 1 \end{bmatrix}.$$

Similarly, the matrix of Ω_m is

$$\begin{bmatrix} \text{ref } \phi & 0 \\ 0 & 1 \end{bmatrix},$$

and, hence, $\Omega_\ell \Omega_m$ has the matrix.

$$\begin{bmatrix} \text{rot } 2(\theta - \phi) & 0 \\ 0 & 1 \end{bmatrix}.$$

We use the same definition for rotation as in $\mathbf{E}^2$.

Definition. *If α and β are lines passing through a point P, then the isometry $\Omega_\alpha \Omega_\beta$ is called a* rotation *about P. The special case $\alpha = \beta$ determines the identity, a trivial rotation. If $\alpha \neq \beta$, the rotation is said to be nontrivial. We denote the set of all rotations about P by $\text{ROT}(P)$. Note that $\text{ROT}(P) \cong$* **SO(2)**.

The above calculations reduce the algebra of rotations about a point P to that used in $\mathbf{E}^2$. Thus, it is easy to verify the following important results concerning rotations of $\mathbf{S}^2$.

Theorem 15 (Three reflections theorem). *Let α, β, and γ be three lines through a point P. Then there is a unique line δ through P such that*

$$\Omega_\alpha \Omega_\beta \Omega_\gamma = \Omega_\delta.$$

Theorem 16 (Representation theorem for rotations). *Let $T = \Omega_\alpha \Omega_\beta$ be any member of $\text{ROT}(P)$, and let ℓ be any line through P. Then there exist unique lines m and m' through P such that*

$$T = \Omega_\ell \Omega_m = \Omega_{m'} \Omega_\ell.$$

Definition. *Let ℓ be any line, and let m and n be perpendicular to ℓ. The* *transformation $\Omega_m \Omega_n$ is called a* translation *along ℓ. If $m \neq n$, the translation is said to be* nontrivial.

Remark: Unlike the Euclidean plane, $\mathbf{S}^2$ does not have parallel lines. In fact, if two lines are perpendicular to ℓ, then they intersect in the poles of ℓ. Thus, our study of products of two reflections simplifies tremendously.

Theorem 17.
  i. *Every translation of $\mathbf{S}^2$ is also a rotation.*
  ii. *Every rotation of $\mathbf{S}^2$ is also a translation.*

The translations along a line ℓ in $\mathbf{E}^2$ were parametrized by the real numbers. If we take into account the periodic nature of our parametrization of lines of $\mathbf{S}^2$, we can obtain the analogous relationships among reflections in lines perpendicular to ℓ.

Consider now two lines α and β perpendicular to ℓ. Let P be an arbitrary point of ℓ. Let ξ be a pole of ℓ, and let $Q = \xi \times P$. Then we may choose numbers a and b such that

$$(\cos a)P + (\sin a)Q \in \alpha, \quad (\cos b)P + (\sin b)Q \in \beta.$$

Then it is easy to check that $(-\sin a)P + (\cos a)Q$ is a pole of α, and $(-\sin b)P + (\cos b)Q$ is a pole of β. As we observed in dealing with rotations,

$$\Omega_\alpha = \begin{bmatrix} \cos 2a & \sin 2a & 0 \\ \sin 2a & -\cos 2a & 0 \\ 0 & 0 & 1 \end{bmatrix}$$

with respect to the orthonormal basis $\{P, Q, \xi\}$, and, thus

$$\Omega_\alpha \Omega_\beta = \begin{bmatrix} \cos 2(a - b) & -\sin 2(a - b) & 0 \\ \sin 2(a - b) & \cos 2(a - b) & 0 \\ 0 & 0 & 1 \end{bmatrix}$$

$$= \begin{bmatrix} \text{rot } 2(a - b) & 0 \\ 0 & 1 \end{bmatrix}.$$

We denote the set of all translations along ℓ by TRANS (ℓ). The group generated by all reflections in the pencil $\mathscr{P}$ of lines perpendicular to ℓ is denoted by REF($\mathscr{P}$).

Theorem 18. *TRANS(ℓ) is an abelian group that coincides with ROT(ξ), where ξ is a pole of ℓ.*

As a consequence of the preceding discussion, we can assert and interpret the following theorems.

Theorem 19 (Three reflections theorem).   *Let* α, β, *and* γ *be three lines of a pencil* $\mathscr{P}$ *with common perpendicular* ℓ. *Then there is a unique fourth line* δ *of this pencil such that*

$$\Omega_\alpha\Omega_\beta\Omega_\gamma = \Omega_\delta.$$

Theorem 20 (Representation theorem for translations).   *Let* $T = \Omega_\alpha\Omega_\beta$ *be any member of* $\mathrm{TRANS}(\ell)$. *If* m *and* n *are arbitrary lines perpendicular to* ℓ, *there exist unique lines* m' *and* n' *such that*

$$T = \Omega_m\Omega_{m'} = \Omega_{n'}\Omega_n.$$

Corollary.   *Every element of* $\mathrm{REF}(\mathscr{P})$ *is either a translation along* ℓ *or a reflection in a line of* $\mathscr{P}$.

Definition.   *If* α *and* β *are lines perpendicular to a line* ℓ, *then* $\Omega_\alpha\Omega_\beta\Omega_\ell$ *is called a* glide reflection *with axis* ℓ.

Remark: If $\{e_1, e_2, e_3\}$ is an orthonormal basis with e_3 a pole of ℓ, then

$$\Omega_\ell e_1 = e_1, \quad \Omega_\ell e_2 = e_2, \quad \Omega_\ell e_3 = -e_3.$$

Corollary.   *With respect to an orthonormal basis of this type, a glide reflection with axis* ℓ *has the form*

$$\begin{bmatrix} \cos\lambda & -\sin\lambda & 0 \\ \sin\lambda & \cos\lambda & 0 \\ 0 & 0 & -1 \end{bmatrix}.$$

Definition.   *An isometry that is a product of a finite number of reflections is called a* motion.

   Simplifying the proofs of Theorems 35–37 of Chapter 1 to take into account the absence of parallelism yields a proof of the following basic structure theorem for motions.

Theorem 21.   *Every motion is the product of two or three suitably chosen reflections.*

Orthogonal transformations of $\mathbf{E}^3$

Definition.   *A mapping* $T: \mathbf{E}^3 \to \mathbf{E}^3$ *is said to be* orthogonal *if*
  i.   *T is linear.*
  ii. $\langle Tx, Ty \rangle = \langle x, y \rangle$ *for all* $x, y \in \mathbf{E}^3$.

Theorem 22. *A linear mapping T is orthogonal if and only if its matrix A (with respect to some orthonormal basis) satisfies $A^tA = I$.*

Proof: Suppose that T is orthogonal and that $\{e_i\}$ is orthonormal. Then

$$\langle Te_i, Te_j \rangle = \left\langle \sum_{k=1}^{3} a_{ki}e_k, \sum_{\ell=1}^{3} a_{\ell j}e_\ell \right\rangle$$

$$= \sum_{k,\ell=1}^{3} a_{ki}a_{\ell j}\langle e_k, e_\ell \rangle$$

$$= \sum_{k=1}^{3} a_{ki}a_{kj} = (A^tA)_{ij}.$$

But $\langle Te_i, Te_j \rangle = \langle e_i, e_j \rangle$. Thus, $A^tA = I$.

Conversely, if $A^tA = I$, with respect to some orthonormal basis, the same calculations show that

$$\langle Te_i, Te_j \rangle = \langle e_i, e_j \rangle.$$

Thus, for any $x = \Sigma x_ie_i$ and $y = \Sigma y_je_j$, the linearity of the inner product yields $\langle Tx, Ty \rangle = \langle x, y \rangle$. $\square$

Definition. *A 3×3 matrix satisfying $A^tA = I$ is called an* orthogonal matrix.

Remark: If A is an orthogonal matrix, then $A^{-1} = A^t$, so that $AA^t = I$ also. Finally, $\det(A^tA) = (\det A)^2 = 1$, so that $\det A = \pm 1$.

Theorem 23.
i. *The set $\mathbf{O}(3)$ of all orthogonal transformations of $\mathbf{E}^3$ is a group called the* orthogonal group of $\mathbf{R}^3$.
ii. *The set $\mathbf{SO}(3)$ of orthogonal transformations with determinant $+1$ is a subgroup of $\mathbf{O}(3)$ called the* special orthogonal group.

Proof: To prove (i), we check that the set of orthogonal matrices is closed under multiplication and the taking of inverses. First, note that if A and B are orthogonal, then

$$(AB)^tAB = B^tA^tAB = B^tB = I,$$

so that

$$(AB)^t = (AB)^{-1}.$$

Next, if A is orthogonal, then

$$(A^{-1})^tA^{-1} = (A^t)^tA^t = AA^t = I,$$

97

so that A^{-1} is orthogonal. This completes the proof of (i).

To prove (ii), note that $A, B \in \mathbf{SO(3)}$ gives $\det(AB) = (\det A)(\det B) = 1 \cdot 1 = 1$, so that the product AB is in $\mathbf{SO(3)}$. Secondly, if $A \in \mathbf{SO(3)}$, then

$$\det(A^{-1}) = 1/\det A$$
$$= 1/1$$
$$= 1,$$

so that A^{-1} is also in $\mathbf{SO(3)}$. □

Euler's theorem

Earlier, we observed that every reflection and hence every motion may be regarded as arising from an orthogonal transformation. In this section we show the converse – every orthogonal transformation induces a motion of $\mathbf{S}^2$.

It turns out that $\mathbf{SO(3)}$ corresponds precisely to the set of rotations whereas orthogonal transformations with determinant -1 correspond to reflections and glide reflections.

We first prove the following theorem of Euler.

Theorem 24. *For each T in $\mathbf{SO(3)}$ there is an x in $\mathbf{S}^2$ such that $Tx = x$.*

Proof: We begin by trying to solve an apparently harder problem. We attempt to find all nonzero vectors x in $\mathbf{R}^3$ such that Tx and x are proportional. This means that we must solve the equation

$$Tx = \lambda x;$$

that is, $(T - \lambda I)x = 0$ for some real number λ.

We know from Appendix D that a nontrivial solution would require that λ satisfy $\det(T - \lambda I) = 0$. The expression on the left is a polynomial of degree 3 in λ called the *characteristic polynomial.* Write

$$\mathrm{char}(t) = \det(T - tI).$$

There are two possibilities for factoring the polynomial:
1. $\mathrm{char}(t) = (\lambda_1 - t)(\lambda_2 - t)(\lambda_3 - t)$ (three real roots).
2. $\mathrm{char}(t) = (\lambda - t)(\mu - t)(\bar{\mu} - t)$ (one real and two complex conjugate roots).

In either case there is, for each real root λ, a unit vector x such that $Tx = \lambda x$. In case (1)

$$1 = |Tx|^2 = \lambda_i^2|x|^2 = \lambda_i^2,$$

so that $\lambda_i = \pm 1$. On the other hand, the product of the roots must be equal to the determinant of T, and, hence, at least one of the roots must be $+1$.

In the second case the product of the roots $\lambda\mu\bar{\mu} = \lambda|\mu|^2$ must again be $+1$. Therefore, $\lambda = +1$. $\square$

Corollary. *For any $T \in \mathbf{SO(3)}$ the restriction of T to $\mathbf{S}^2$ is a rotation.*

Proof: By Euler's theorem there is a point of $\mathbf{S}^2$ that is mapped to itself by T. Choose an orthonormal basis $\{e_1, e_2, e_3\}$ with $Te_3 = e_3$. Then for suitable choice of θ,

$$
\begin{aligned}
Te_1 &= (\cos\theta)e_1 + (\sin\theta)e_2, \\
Te_2 &= \pm((-\sin\theta)e_1 + (\cos\theta)e_2), \\
Te_3 &= e_3
\end{aligned}
$$

by the same argument we used in the lemma to Theorem 38, Chapter 1. Because $\det T = 1$, the positive sign should be used in Te_2, and, hence, the matrix of T is

$$
\begin{bmatrix}
\cos\theta & -\sin\theta & 0 \\
\sin\theta & \cos\theta & 0 \\
0 & 0 & 1
\end{bmatrix}.
$$

Because T can be factored into the product of two reflections, it is clearly a rotation. $\square$

Remark: Because every rotation arises from the product of two orthogonal transformations, it must be the restriction of a member of $\mathbf{SO(3)}$.

For all practical purposes $\mathbf{SO(3)}$ may be identified with the set of all rotations. In the future we will frequently use "is" when we really mean "arises from" or "is the restriction of," leaving it to the reader to make the distinctions when necessary.

Not all orthogonal transformations are rotations, of course.

Definition. *The antipodal map E is the transformation defined by*

$$Ex = -x.$$

With respect to any orthonormal basis, E has the matrix

$$
\begin{bmatrix}
-1 & 0 & 0 \\
0 & -1 & 0 \\
0 & 0 & -1
\end{bmatrix}.
$$

The antipodal map is a glide reflection because it can be factored

$$
\begin{bmatrix}
-1 & 0 & 0 \\
0 & 1 & 0 \\
0 & 0 & 1
\end{bmatrix}
\begin{bmatrix}
1 & 0 & 0 \\
0 & -1 & 0 \\
0 & 0 & 1
\end{bmatrix}
\begin{bmatrix}
1 & 0 & 0 \\
0 & 1 & 0 \\
0 & 0 & -1
\end{bmatrix}
$$

as the product of the three reflections in lines whose poles form an orthonormal basis. (Three lines of this type are said to form a *self-polar triangle*. Each vertex is a pole of the opposite side.) However, this glide reflection does not have a unique axis. Every line is an axis.

The antipodal map E is a convenient tool for analyzing orthogonal transformations.

Theorem 25. *Every orthogonal transformation restricts to a motion of* $\mathbf{S}^2$.

Proof: Let T be orthogonal. If $\det T = 1$, then T is a rotation and, thus, a motion. If $\det T = -1$, then ET is a rotation ρ. Thus, $T = E\rho$. (Note that E is its own inverse.) This exhibits T as a motion. □

Isometries

Definition. *A function* $T: \mathbf{S}^2 \to \mathbf{S}^2$ *is an* isometry *if*

$$d(Tx, Ty) = d(x, y)$$

for all x, y *in* $\mathbf{S}^2$.

We recall that every reflection and hence every motion is an isometry. Further, we recall that every orthogonal transformation restricts to a motion of $\mathbf{S}^2$. Because each orthogonal transformation is determined by its value on unit vectors, each isometry is the restriction of at most one orthogonal transformation. We now announce the major result of this section.

Theorem 26. *For every isometry* T_0 *of* $\mathbf{S}^2$ *there is an orthogonal transformation* T *coinciding with* T_0 *on* $\mathbf{S}^2$.

Proof: Let $\{e_1, e_2, e_3\}$ be any orthonormal basis of $\mathbf{E}^3$. Because T_0 is an isometry, we have $\langle T_0 e_i, T_0 e_j \rangle = \langle e_i, e_j \rangle$. Each point of $\mathbf{E}^3$ is of the form λx for some $x \in \mathbf{S}^2$ and $\lambda \geq 0$. Define $T: \mathbf{E}^3 \to \mathbf{E}^3$ by

$$T(\lambda x) = \lambda T_0 x \quad \text{if} \quad \lambda x \neq 0,$$

$$T(0) = 0.$$

We must now check that T is orthogonal. First we deal with linearity. For any $x \in \mathbf{S}^2$, $Tx = T_0 x$ and

$$Tx = \sum \langle Tx, Te_i \rangle Te_i = \sum \langle x, e_i \rangle Te_i$$

because $\{Te_i\}$ is also an orthonormal basis. Thus,

$$T(\lambda x) = \lambda T_0 x = \lambda Tx = \lambda \sum \langle x, e_i \rangle Te_i$$
$$= \sum \langle \lambda x, e_i \rangle Te_i.$$

In other words, for any $u \neq 0$ in $\mathbf{E}^3$ (and also more obviously for $u = 0$) we have

$$Tu = \sum \langle u, e_i \rangle Te_i.$$

This expression is clearly linear in u. Furthermore, if v is another vector in $\mathbf{E}^3$,

$$\langle Tu, Tv \rangle = \sum \langle u, e_i \rangle \langle v, e_j \rangle \langle Te_i, Te_j \rangle$$
$$= \sum \langle u, e_i \rangle \langle v, e_i \rangle = \langle u, v \rangle,$$

so that T is orthogonal. □

Fixed points and fixed lines of isometries

We now characterize the various types of isometries according to the nature of their sets of fixed points and fixed lines.

Theorem 27.
i.   *A nontrivial rotation has exactly two antipodal fixed points.*
ii. *A reflection has a line of fixed points – its axis.*
iii. *A glide reflection has no fixed points.*
iv. *The identity leaves all points fixed.*

Theorem 28.   *An isometry T leaves a line with pole ξ fixed if and only if $T\xi = \pm\xi$.*

Theorem 29.
i.   *A half-turn leaves fixed lines all through a point (the center) and their common perpendicular.*
ii. *A nontrivial rotation other than a half-turn leaves fixed only the polar line of its center.*
iii. *A reflection leaves fixed its axis and all lines perpendicular to it (same fixed lines as half-turn).*
iv. *A glide reflection other than the antipodal map leaves only its axis fixed.*
v.   *The antipodal map and the identity leave every line fixed.*

Further representation theorems

Because the rotations constitute a subgroup of $\mathcal{I}(\mathbf{S}^2)$, it is clear that the product of two successive half-turns is a rotation. It may surprise you to learn that the converse is also true. Specifically, we have the following representation theorem for rotations:

Theorem 30. *Every rotation can be written as the product of two half-turns.*

Proof: Let T be a rotation. By Theorem 16 there are lines ℓ and m such that $T = \Omega_\ell \Omega_m$. Let n be a common perpendicular to ℓ and m, meeting ℓ and m in points P and Q, respectively. Then

$$T = \Omega_\ell \Omega_n \Omega_n \Omega_m = H_P H_Q. \qquad \square$$

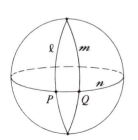

Figure 4.13 $\Omega_\ell \Omega_m = H_P H_Q$.

Remark: In $\mathbf{E}^2$ the product of two half-turns is a translation, and every translation can be so represented. Theorem 30 points out an instance in which it is more productive to think of elements of $\mathbf{SO(3)}$ as translations rather than rotations. The construction is illustrated in Figure 4.13.

Theorem 31. *Let P be a point of $\mathbf{S}^2$. Then for all $x \in \mathbf{S}^2$,*

$$H_P x = -x + 2\langle x, P \rangle P.$$

Proof: Let ξ and η be poles of perpendicular lines through P. Then $\{\xi, \eta, P\}$ is an orthonormal basis, and the identity

$$x = \langle x, P \rangle P + \langle x, \xi \rangle \xi + \langle x, \eta \rangle \eta$$

holds on $\mathbf{S}^2$. Now it is a straightforward calculation that

$$H_P x = x - 2\langle x, \xi \rangle \xi - 2\langle x, \eta \rangle \eta.$$

Putting these two results together yields the desired expression for $H_P x$. $\square$

Corollary. *Let P be any point of $\mathbf{S}^2$, and let ℓ be its polar line. Then*

$$\Omega_\ell H_P = H_P \Omega_\ell = E,$$

where E is the antipodal map.

Proof: The result of Theorem 31 may be written

$$H_P x = -x - 2\langle -x, P \rangle P = -(x - 2\langle x, P \rangle P)$$

for all $x \in \mathbf{S}^2$. In other words,

$$H_P = \Omega_\ell E = E \Omega_\ell;$$

that is,

$$\Omega_\ell H_P = H_P \Omega_\ell = E. \qquad \square$$

We have shown that the antipodal map may be represented as the product of a half-turn and a reflection. We have also determined precisely what combinations can occur in the representation. We now do the same for arbitrary glide reflections.

Let T be a glide reflection other than the antipodal map E. Suppose that its axis ℓ has a point P as a pole. We speak of P as a *pole of T*.

Theorem 32.
i. *For each line m through P there is a unique point Q such that $\Omega_m H_Q = T$. The point Q will necessarily lie on ℓ.*
ii. *For each point Q on ℓ, there is a unique line m such that $\Omega_m H_Q = T$. The line m will necessarily pass through P.*

Remark: For any line m and any point Q, $\Omega_m H_Q$ is a glide reflection whose axis is perpendicular to m and passes through Q.

Segments

In spherical geometry the notion of betweenness is ambiguous. Given any three collinear points, it is possible to regard any one as being between the other two.

On the other hand, a choice of two points on a line ℓ induces a decomposition of ℓ into two subsets that behave much like segments do in $\mathbf{E}^2$. We adopt the following definition.

Definition. *A subset s of $\mathbf{S}^2$ is called a* segment *if there exist points P and Q with $\langle P, Q \rangle = 0$ and numbers $t_1 < t_2$ with $t_2 - t_1 < 2\pi$ such that*

$$s = \{(\cos t)P + (\sin t)Q | t_1 \leq t \leq t_2\}.$$

Remark: All points of a segment are collinear. Each segment determines a unique line. On the other hand, a segment does not determine the data P, Q, t_1, t_2 uniquely. In fact, we have

Theorem 33. *Let s be a segment determined (as in the definition) by P, Q, t_1, t_2 and also by $\tilde{P}$, $\tilde{Q}$, $\tilde{t}_1$, $\tilde{t}_2$. Then*
i. $t_2 - t_1 = \tilde{t}_2 - \tilde{t}_1$. *This number is called the* length *of the segment.*
ii. *If we write $\alpha(t) = (\cos t)P + (\sin t)Q$ and $\tilde{\alpha}(t) = (\cos t)\tilde{P} + (\sin t)\tilde{Q}$, we have*

$$\{\alpha(t_1), \alpha(t_2)\} = \{\tilde{\alpha}(\tilde{t}_1), \tilde{\alpha}(\tilde{t}_2)\}.$$

These points are called the end *points of s. All other points of s are called* interior points *of s.*
iii. $P \times Q = \pm \tilde{P} \times \tilde{Q}$.
These points are the poles of the line on which s lies.

Proof: First, note that replacing $\tilde{Q}$ by its negative, and $[\tilde{t}_1, \tilde{t}_2]$ by $[-\tilde{t}_2, -\tilde{t}_1]$ do not change conditions (i)–(iii). This allows us to assume that there is a number ϕ such that $\tilde{P} = \alpha(\phi)$ and $\tilde{Q} = \alpha(\phi + \pi/2)$. A short computation now reveals that

$$\tilde{\alpha}(u) = (\cos (u + \phi))P + (\sin (u + \phi))Q = \alpha(u + \phi),$$

and, hence, that $\alpha([t_1 + \phi, t_2 + \phi]) = \alpha([t_1, t_2])$. A fundamental property of the trigonometric functions (see the lemma in Appendix F) shows that the two intervals are translates of each other mod 2π. In particular, they have the same length and end points. Finally,

$$\tilde{P} \times \tilde{Q} = ((\cos \phi)P + (\sin \phi)Q) \times ((-\sin \phi)P + (\cos \phi)Q)$$
$$= (\cos^2 \phi + \sin^2 \phi)(P \times Q) = P \times Q. \qquad \square$$

Corollary. *Let A and B be arbitrary points satisfying $\langle A, B \rangle = 0$. Let ∂ be any segment lying on $\overleftrightarrow{AB}$. Then there is a unique interval $[a, b]$ such that $0 \leqslant a < 2\pi$ and*

$$\partial = \{(\cos t)A + (\sin t)B | a \leqslant t \leqslant b\}.$$

Proof: First represent ∂ as

$$\{(\cos t)P + (\sin t)Q | t_1 \leqslant t \leqslant t_2\},$$

where

$$P = (\cos \phi)A + (\sin \phi)B \quad \text{and} \quad Q = (-\sin \phi)A + (\cos \phi)B$$

for some number ϕ. (This is possible because ∂ is a segment.) Clearly,

$$(\cos t)P + (\sin t)Q = \cos(t + \phi)A + \sin(t + \phi)B$$

for all real t. Thus, we should choose $a \equiv t_1 + \phi$ mod 2π in the interval $[0, 2\pi)$ and $b = a + (t_2 - t_1)$. $\qquad \square$

Theorem 34. *Let A and B be nonantipodal points. Then there are exactly two segments having A and B as end points. Their union is the line $\overleftrightarrow{AB}$, and their intersection is the set $\{A, B\}$.*

Proof: Let ξ be a unit vector in the direction $[A \times B]$, and set $Q = \xi \times A$. Then there is a unique number $L \in (0, 2\pi)$ such that $B = (\cos L)A + (\sin L)Q$. The segments

$$\{(\cos t)A + (\sin t)Q | 0 \leqslant t \leqslant L\}$$

and

$$\{(\cos t)A - (\sin t)Q | 0 \leqslant t \leqslant 2\pi - L\}$$

have A and B as end points. Because the second segment may be rewritten

$$\{(\cos t)A + (\sin t)Q \mid 0 \geqslant t \geqslant L - 2\pi\},$$

we see that the union of the segments is $\overleftrightarrow{AB}$. The same observation shows that the two segments have no interior points in common and thus intersect only at their end points. □

Definition. *Let A and B be nonantipodal points. The longer of the two segments having A and B as end points is called the* major segment *AB. The shorter one is called the* minor segment *AB. The two segments are said to be* complements *of each other and may be referred to as* complementary *segments. See Figure 4.14.*

Definition. *If A and B are antipodal points, each of the (infinitely many) segments having A and B as end points is called a* half-line.

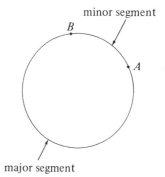

Figure 4.14 Complementary segments.

Remark: The length of a minor segment AB is $d(A, B)$. The length of a major segment AB is $2\pi - d(A, B)$. The length of a half-line is π.

Remark: In spherical geometry it is ambiguous to speak of the segment AB without specifying whether we mean the major or the minor segment.

Theorem 35. *Let P, Q, and X be distinct points of* $\mathbf{S}^2$. *If P and Q are not antipodal, a point X lies on the minor segment PQ if and only if*

$$d(P, X) + d(X, Q) = d(P, Q). \qquad (4.1)$$

Proof: Choose $\tilde{P}$ (orthogonal to P) so that the segment in question is

$$\{(\cos t)P + (\sin t)\tilde{P} \mid 0 \leqslant t \leqslant L\},$$

where $L = d(P, Q)$.

Assume first that X lies on the segment. Then we may write

$$X = (\cos \phi)P + (\sin \phi)\tilde{P}, \qquad (4.2)$$

where $\phi \in (O, L)$. We now verify (4.1) by computing

$$d(P, X) = \cos^{-1} \cos \phi = \phi$$

and

$$d(X, Q) = \cos^{-1} \cos (L - \phi) = L - \phi.$$

Conversely, suppose that (4.1) holds. Then by the corollary to Theorem 8, X is on the line $\overleftrightarrow{PQ}$, so the representation (4.2) holds for a unique number $\phi \in (0, 2\pi)$. It can be verified (Exercise 33) that $L < \phi < 2\pi$ contradicts (4.1). The alternative, $0 < \phi < L$, must therefore hold, and X must lie on the minor segment PQ. □

105

Remark: If P and Q are antipodal, the identity (4.1) holds automatically for all X on $\mathbf{S}^2$.

Theorem 36. *Let P and Q be antipodal points. Let R be any other point. Then the union of the minor segments PR and RQ is a half-line. If $R' = -R$, the union of the four minor segments PR, RQ, PR', and $R'Q$ is the line $\overleftrightarrow{PR}$.*

Proof: Let $\tilde{P}$ be a point on $\overleftrightarrow{PR}$ orthogonal to P such that

$$R = (\cos L)P + (\sin L)\tilde{P},$$

where $L = d(P, R)$. Then the minor segments in question are

$$PR = \{(\cos t)P + (\sin t)\tilde{P} \mid 0 \leqslant t \leqslant L\},$$
$$RQ = \{(\cos t)P + (\sin t)\tilde{P} \mid L \leqslant t \leqslant \pi\},$$
$$PR' = \{(\cos t)P + (\sin t)(-\tilde{P}) \mid 0 \leqslant t \leqslant \pi - L\},$$
$$R'Q = \{(\cos t)P + (\sin t)(-\tilde{P}) \mid \pi - L \leqslant t \leqslant \pi\}.$$

But we may rewrite PR' and $R'Q$ as

$$PR' = \{(\cos t)P + (\sin t)\tilde{P} \mid -(\pi - L) \leqslant t \leqslant 0\},$$
$$R'Q = \{(\cos t)P + (\sin t)\tilde{P} \mid -\pi \leqslant t \leqslant -(\pi - L)\}.$$

Now

$$\overleftrightarrow{PR} = \{(\cos t)P + (\sin t)\tilde{P} \mid -\pi \leqslant t \leqslant \pi\}$$

is the union of the four segments. $\qquad\square$

Theorem 37. *Let T be an isometry. Then*
i.   *If a is a minor segment, so is $T\mathit{a}$.*
ii. *If a is a half-line, so is $T\mathit{a}$.*
iii. *If a is a major segment, so is $T\mathit{a}$.*

Proof:
i.   Let a be the minor segment AB. Then by Theorem 35

$$\mathit{a} = \{X \mid d(A, X) + d(X, B) = d(A, B)\}.$$

Thus,

$$\begin{aligned}
T\mathit{a} &= \{TX \mid d(A, X) + d(X, B) = d(A, B)\} \\
&= \{TX \mid d(TA, TX) + d(TX, TB) = d(TA, TB)\} \\
&= \{Y \mid d(TA, Y) + d(Y, TB) = d(TA, TB)\}.
\end{aligned}$$

Again applying Theorem 35, we see that $T\mathit{a}$ is a minor segment with end points TA and TB.

ii. Let $\mathfrak{s}$ be a half-line lying on a line ℓ and having end points A and B. Let C be any other point of $\mathfrak{s}$. Then $\mathfrak{s}$ is the union of the minor segments AC and CB. Thus, $T\mathfrak{s}$ is the union of the minor segments with end point sets $\{TA, TC\}$ and $\{TC, TB\}$. Furthermore, TA and TB are antipodal. We conclude that $T\mathfrak{s}$ is a half-line.

iii. Suppose that A is a major segment with end points A and B. We know that T takes the minor segment AB to the minor segment with end points TA and TB. Because T takes the line $\ell = \overleftrightarrow{AB}$ to the line through TA and TB, it must take the complementary segment to the corresponding major segment on $T\ell$. □

Rays, angles, and triangles

In spherical geometry we define a *ray* to be a half-line with one end point removed. The other end point is called the *origin* of the ray.

Suppose that PQ is a minor segment of length L represented in the standard way by

$$\{(\cos t)P + (\sin t)\tilde{P}|0 \leq t \leq L\}.$$

Then

$$\overrightarrow{PQ} = \{(\cos t)P + (\sin t)\tilde{P}|0 \leq t < \pi\}$$

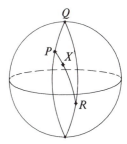

Figure 4.15 X is in the interior of $\measuredangle PQR$.

is the unique ray through Q with origin P.

We can define angle just as we did in $\mathbf{E}^2$. In this case a straight angle (as a set of points) is just a line with one point removed.

Definition. *Let $\measuredangle PQR$ be an angle. A point X is in the interior of the angle if the minor segment XP does not intersect $\overleftrightarrow{QR}$ and the minor segment XR does not intersect $\overleftrightarrow{QP}$.*

The set of points in the interior of an angle is called a *lune*. A pair of distinct lines decomposes $\mathbf{S}^2$ into four lunes.

Remark: Each line ℓ decomposes $\mathbf{S}^2$ into two half-planes. Half-planes may be defined as in $\mathbf{E}^2$, except that the segments used in the definition are minor segments. A lune is the intersection of two half-planes. These ideas are developed further in Exericse 35. See also Figures 4.15 and 4.16.

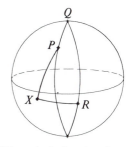

Figure 4.16 X is not in the interior of $\measuredangle PQR$.

Because rays no longer have direction vectors, we must find another way of defining the radian measure of an angle. Thinking in $\mathbf{E}^3$ for the moment, we see that the vectors that are poles of two intersecting lines are unit normals to these lines. Thus, the angle between the lines corresponds to the angle between these unit normals.

107

However, when two lines intersect, they determine four angles. How to choose the right sign when computing the radian measure of an angle is not as intuitively clear. The correct definition is the following:

Definition. *The radian measure of an angle $\angle PQR$ is*

$$\cos^{-1}\left\langle \frac{Q \times P}{|Q \times P|}, \frac{Q \times R}{|Q \times R|}\right\rangle.$$

Let P, Q, and R be three noncollinear points. The *triangle PQR* is defined to be the union of the three minor segments PQ, QR, and PR. The segments are called *sides* of the triangle, and the length of each side is equal to the distance between its end points.

The interior of a triangle is defined as in $\mathbf{E}^2$.

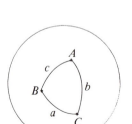

Figure 4.17 A triangle in spherical geometry, first view.

Remark: Our definition of triangle is not the only possible one. However, it is the easiest to work with because it has the following properties:

i. Three noncollinear points determine a triangle.

ii. Every triangle lies in some half-plane.

Spherical trigonometry

Let ABC be a triangle. Let a be the length of the side BC, b the length of AC, and c the length of AB. See Figures 4.17 and 4.18. Note that

$$|B \times C|^2 = |B|^2|C|^2 - \langle B, C\rangle^2 = 1 - \cos^2 a = \sin^2 a$$

and

$$\langle A \times B, A \times C\rangle = \langle B, C\rangle - \langle A, C\rangle\langle A, B\rangle$$
$$= \cos a - \cos b \cos c. \tag{4.3}$$

Hence, we have the spherical version of the Law of Cosines:

$$\cos A = \frac{\cos a - \cos b \cos c}{\sin b \sin c},$$

where we have written A as an abbreviation for the radian measure of $\angle BAC$.

Now

$$1 - \cos A = \frac{\cos(b - c) - \cos a}{\sin b \sin c} = 2\sin^2 \frac{A}{2}$$

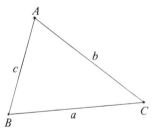

Figure 4.18 A triangle in spherical geometry, second view.

and

$$1 + \cos A = \frac{\cos a - \cos(b + c)}{\sin b \sin c} = 2\cos^2 \frac{A}{2}.$$

Thus

$$\sin^2 A = 4 \sin^2 \frac{A}{2} \cos^2 \frac{A}{2} = \frac{k}{\sin^2 b \, \sin^2 c},$$

where k is the product of the two factors

$$-2 \sin \frac{b - c + a}{2} \sin \frac{b - c - a}{2} \quad \text{and} \quad -2 \sin \frac{a + b + c}{2} \sin \frac{a - b - c}{2}.$$

Putting $a + b + c = 2s$, we have

$$\frac{\sin^2 A}{\sin^2 a} = \frac{4 \sin s \, \sin (s - a) \, \sin (s - b) \, \sin (s - c)}{\sin^2 a \, \sin^2 b \, \sin^2 c},$$

and, hence,

$$\frac{\sin A}{\sin a} = \frac{2(\sin s \, \sin(s - a) \, \sin(s - b) \, \sin(s - c))^{1/2}}{\sin a \, \sin b \, \sin c}. \qquad (4.4)$$

Note that the right side is symmetric in a, b, and c, so that we may conclude that

$$\frac{\sin A}{\sin a} = \frac{\sin B}{\sin b} = \frac{\sin C}{\sin c}.$$

This is the Law of Sines for spherical trigonometry.

There is a nice relationship between angles and sides of a spherical triangle that arises from the pole–polar correspondence. Each of the formulas we have developed has a counterpart with the roles of angle and side interchanged. In particular, we have the two versions of the Law of Cosines. The first was proved earlier in this section. The second is Exercise 37.

Theorem 38. *In the notation of this section we have*

i. $\cos A = \dfrac{\cos a - \cos b \, \cos c}{\sin b \, \sin c},$

ii. $\cos a = \dfrac{\cos A + \cos B \, \cos C}{\sin B \, \sin C}.$

$$(4.5)$$

Corollary. *The lengths of the sides of a triangle are completely determined by the radian measures of its angles.*

Rectilinear figures

We may define rectilinear figures as in Euclidean geometry. All the definitions and proofs are either exactly the same or very similar. In particular, every symmetry of a rectilinear figure permutes the vertices. The vertices of a complete rectilinear figure on $\mathbf{S}^2$ occur in antipodal pairs. Thus, every symmetry also permutes the set of antipodal pairs of vertices.

In particular, let Δ be a triangle with vertices PQR. Let G be the stabilizer of P in $\mathscr{S}(\Delta)$. Now G consists of rotations about P and reflections in lines through P. Every member of G must permute the set $\{Q, R\}$. If P lies on the perpendicular bisector m of QR, then Ω_m will be in G. However, if $\ell = \overleftrightarrow{QR}$, Ω_ℓ will not be in G. Thus, $G = \{I\}$ or $G = \{I, \Omega_m\}$. Also, the orbit of P consists of at most three elements P, Q, and R. Thus, $\#\mathscr{S}(\Delta) \leqslant 3 \times 2 = 6$. If Δ is isosceles, then $\#\mathscr{S}(\Delta) = 2$. If Δ is equilateral, then $\#\mathscr{S}(\Delta) = 6$. If Δ is scalene, $\#\mathscr{S}(\Delta) = 1$.

These arguments show that as far as symmetry is concerned, spherical triangles behave just like Euclidean triangles.

Theorem 39. *Let $\mathscr{F}$ be a rectilinear figure having at least three noncollinear vertices. Then $\mathscr{S}(\mathscr{F})$ is a finite group.*

Proof: Every symmetry of $\mathscr{F}$ induces a permutation on the vertices. But a linear transformation is determined by its action on three linearly independent vectors because they form a basis (Theorem 4D). Therefore, there is at most one isometry realizing each permutation of the vertices of $\mathscr{F}$. □

Theorem 40. *Let $\mathscr{F}$ be the complete rectilinear figure consisting of three mutually perpendicular lines. Then $\mathscr{S}(\mathscr{F})$ is a group of order 48.*

Proof: Let P, Q, and R be poles of the three lines (Figure 4.19). Each permutation of the set $\{P, Q, R\}$ determines an isometry whose matrix with respect to the orthonormal basis $\{P, Q, R\}$ is a permutation matrix. (Each row and column has one 1 and two 0's.) For each of the six permutation matrices, there are eight ways of introducing minus signs into the matrix. Each minus sign introduced corresponds to a reflection in one of the lines of the configuration.

Clearly, the 48 matrices so obtained are orthogonal and so define isometries of $\mathbf{S}^2$. The isometries permute the lines of $\mathscr{F}$ and so are symmetries. On the other hand, any symmetry must permute the set of antipodal pairs. It is easy to see that the given constructions yield all such permutations. □

The group $\mathscr{S}(\mathscr{F})$ consists of 24 rotations, 9 reflections, and 15 glide reflections, including the antipodal map.

The figure $\mathscr{F}$ decomposes $\mathbf{S}^2$ into four pairs of antipodal triangles. The symmetries of these triangles provide eight nontrivial rotations. There are also three nontrivial rotations about each of the three antipodal pairs of vertices. There are six half-turns about the midpoints of the segments of the figure. Finally, the identity rounds out our list of 24 rotations.

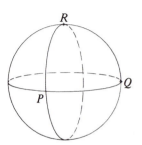

Figure 4.19 Theorem 40. Three mutually perpendicular lines.

Congruence theorems

Now that we have defined some geometrical objects – segments, rays, angles, triangles – we return to our study of the group of isometries of $\mathbf{S}^2$ and how they act on simple figures.

Theorem 41. *Let P and Q be points of $\mathbf{S}^2$. Then there is a unique reflection interchanging them.*

Proof: Let $\xi = (P - Q)/|P - Q|$ and let ℓ be the line whose pole is ξ. Then

$$\Omega_\ell P = P - 2\frac{\langle P - Q, P\rangle(P - Q)}{|P - Q|^2}$$

$$= P - 2\frac{(1 - \langle Q, P\rangle)(P - Q)}{2(1 - \langle Q, P\rangle)}$$

$$= P - (P - Q) = Q.$$

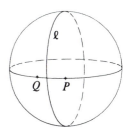

Figure 4.20 The perpendicular bisector of PQ, first view. Ω_ℓ interchanges P and Q.

Thus, Ω_ℓ interchanges P and Q.

To prove uniqueness, suppose that Ω_ℓ and Ω_m are reflections that interchange P and Q. Then the rotation $\Omega_\ell\Omega_m$ leaves both P and Q fixed. If P and Q are not antipodal, then $\Omega_\ell\Omega_m = I$ by Theorem 27, and $\ell = m$. If P and Q are antipodal, then the pole ξ of ℓ satisfies

$$P - 2\langle P, \xi\rangle\xi = -P.$$

Thus, $P = \langle P, \xi\rangle\xi$, and ℓ is the polar line of P. Because the same argument applies to m, we see that $\ell = m$. □

Definition. *Let $\jmath$ be a segment. The perpendicular bisector of $\jmath$ is the unique line ℓ such that Ω_ℓ interchanges the end points of $\jmath$. See Figures 4.20 and 4.21.*

Remark: The perpendicular bisector of $\jmath$ is perpendicular to the line on which $\jmath$ lies.

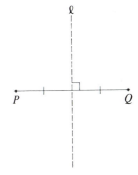

Figure 4.21 The perpendicular bisector of PQ, second view. Ω_ℓ interchanges P and Q.

Theorem 42. *The perpendicular bisector of a segment is the set of all points of $\mathbf{S}^2$ that are equidistant from its end points.*

Proof: This is essentially Exercise 7. □

Definition. *The* midpoint *of a segment is its unique point of intersection with its perpendicular bisector.*

Remark:

i. The midpoint M of a segment $\jmath$ is the unique point of $\jmath$ that is equidistant from the end points of $\jmath$.

111

ii. A segment and its complement have the same perpendicular bisector.

Theorem 42 also suggests a method for constructing an isometry that interchanges two given lines.

Theorem 43. *There are exactly two reflections that interchange a given pair of lines.*

Proof: Choose points P and Q so that the poles of the given lines are $\pm P$ and $\pm Q$. The unique reflection that interchanges P and Q will interchange their polar lines; so will the unique reflection that interchanges P with $-Q$ (and, hence, Q with $-P$).

On the other hand, any reflection that interchanges the lines must send P to Q or P to $-Q$. Thus, the two reflections mentioned are the only ones that interchange the given lines. $\qquad\square$

Theorem 44. *For any angle $\mathcal{A}$ there is a unique reflection that interchanges its arms.*

Proof: We first look at the special cases – the zero angle and the straight angle. In both of these cases any such reflection must leave fixed the line ℓ on which the arms lie and must also have the vertex as a fixed point. Hence, it is either Ω_ℓ or Ω_m, where m is the line through the vertex perpendicular to ℓ. Clearly, Ω_ℓ leaves the zero angle pointwise fixed while Ω_m interchanges the arms of the straight angle.

Now let $\angle PQR$ be an angle that is neither a zero angle nor a straight angle. To simplify the calculation, we may assume that $\langle P, Q \rangle = \langle R, Q \rangle = 0$ and that $P \times R = |P \times R|Q$. (Geometrically, this expresses the fact that $\overleftrightarrow{PQ}$ is the polar line of R.) Let ℓ be the line whose pole is $\xi = (P - R)/|P - R|$. Note that $\langle Q, \xi \rangle = 0$, so that $\Omega_\ell Q = Q$. Also,

$$\Omega_\ell P = P - 2\langle P, \xi \rangle \xi = P - 2\frac{\langle P, P - R \rangle}{|P - R|^2}(P - R)$$
$$= P - (P - R) = R.$$

Similarly, because $|P - R|^2 = 2(1 - \langle P, R \rangle)$, we get that $\Omega_\ell R = Q$, so that Ω_ℓ interchanges the arms of $\mathcal{A}$. $\qquad\square$

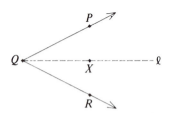

Figure 4.22 Ω_ℓ interchange $\overrightarrow{QP}$ and $\overrightarrow{QR}$.

Definition. *In the notation of Theorem 44 the ray $\overrightarrow{QX}$, where $X = (P + R)/|P + R|$, is called the* bisector *of the angle $\mathcal{A} = \angle PQR$. See Figure 4.22.*

Note that X is a point of ℓ that lies in the interior of $\mathcal{A}$.

Remark: Clearly, Theorem 44 provides a way of proving Theorem 43. Furthermore, if P and Q are any two nonantipodal points, and R is a pole of $\overleftrightarrow{PQ}$, the reflection that interchanges the arms of $\angle PRQ$ will also interchange P and Q. Thus, we can also deduce Theorem 41 from Theorem 44. However, the proofs of Theorems 41 and 43 are easier, and we have decided to include them separately.

We conclude this section with some results on angle addition. The proofs are left to the exercises, and we will not use the results in subsequent theorems. See also Theorem 7.42.

Theorem 45. *If $\overrightarrow{QX}$ is the bisector of an angle $\angle PQR$, then $\angle PQX$ is congruent to $\angle RQX$.*

Theorem 46. *Let $\angle PQR$ be an angle, and let X be a point in its interior. Then the radian measure of $\angle PQR$ is the sum of the radian measures of $\angle PQX$ and $\angle RQX$.*

Remark: In spherical geometry the angle sum for a triangle varies with the size of the triangle. It is easy to check, for example, that if $\{P, Q, R\}$ is an orthonormal basis, then each angle of $\triangle PQR$ is a right angle, so that the sum of the radian measures of the three angles is $3\pi/2$. See Figure 4.23.

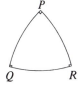

Figure 4.23 A triangle with three right angles.

Theorem 47. *Let P and Q be points on a line ℓ. Then there is exactly one translation along ℓ that takes P to Q.*

Proof: Choose an orthonormal basis $\{e_1, e_2, e_3\}$ such that $e_1 = P$, $Q = (\cos \theta)e_1 + (\sin \theta)e_2$ for some $\theta \in [0, \pi]$, and $e_3 = e_1 \times e_2$ is a pole of ℓ. If $\tau \in \text{TRANS}(\ell)$, there is a number ϕ such that the matrix of τ with respect to $\{e_1, e_2, e_3\}$ is (by Theorem 17)

$$\begin{bmatrix} \cos \phi & -\sin \phi & 0 \\ \sin \phi & \cos \phi & 0 \\ 0 & 0 & 1 \end{bmatrix}$$

If τ takes P to Q, then

$$Q = \tau e_1 = (\cos \phi)e_1 + (\sin \phi)e_2.$$

Thus, $\theta \equiv \phi \bmod 2\pi$, and τ is determined uniquely by this condition. $\quad \square$

Theorem 48. *Let P, Q, P', and Q' be points (not necessarily distinct) lying on a line ℓ. Suppose that $d(P, P') = d(Q, Q')$. Then there is an isometry T such that $TP = Q$ and $TP' = Q'$.*

Proof: By first applying a translation along ℓ, we can arrange that $P = Q$. Then P' and Q' are points of ℓ that are equidistant from P. If $P' = Q'$, we may choose T to be the identity, and we are finished. If not, let $T = \Omega_m$, the reflection that interchanges P' and Q'. Then Ω_m leaves P fixed and sends P' to Q', as required. □

Remark: According to the construction, T could be the identity, a translation, or a reflection.

Theorem 49. *Two segments are congruent if and only if they have the same length.*

Proof: Let $\mathfrak{a}_1$ and $\mathfrak{a}_2$ be congruent minor segments. If T is an isometry taking $\mathfrak{a}_1$ to $\mathfrak{a}_2$, then T takes the end points of $\mathfrak{a}_1$ to those of $\mathfrak{a}_2$ (see proof of Theorem 37). As a result, $\mathfrak{a}_1$ and $T\mathfrak{a}_1 = \mathfrak{a}_2$ have the same length. If two major segments are congruent, so are their complements. Because we know that the complements have equal length, say L, the original major segments must have equal length $\pi - L$.

Conversely, let $\mathfrak{a}_1$ and $\mathfrak{a}_2$ be segments of equal length. We may assume that they are minor segments, because if we find an isometry taking $\mathfrak{a}_1$ to $\mathfrak{a}_2$, it must also relate their complements. First, apply a reflection (Theorem 43) to move $\mathfrak{a}_1$ to the line determined by $\mathfrak{a}_2$. Now translate along this line to make one pair of end points coincide (Theorem 47). If the other pair of end points coincides, we are finished. Otherwise, they are equidistant from the common end point, and the required isometry is completed by applying the reflection that fixes the common end point and interchanges the other two. □

Symmetries of a segment

Theorem 50. *Let $\mathfrak{a}$ be a segment lying on a line ℓ. Let m be its perpendicular bisector, and M its midpoint. Then $\mathcal{S}(\mathfrak{a})$ is the group $\{I, \Omega_\ell, \Omega_m, H_M\}$. Its multiplication table is the same as that in Theorem 2.28.*

Proof: Let $\mathfrak{a}$ be the minor segment PQ. (The major segment has the same symmetries.) It is easy to check that the four given transformations permute the set $\{P, Q\}$ and, hence, are symmetries of $\mathfrak{a}$. (Theorem 37 applies here.) On the other hand, suppose that T is any symmetry of $\mathfrak{a}$. Then T permutes its end points, so that T or $\Omega_m T$ leaves P and Q fixed. By Theorem 27 the only isometries of $\mathbf{S}^2$ leaving P and Q fixed are Ω_ℓ and I. Thus, T must be I, Ω_ℓ, Ω_m, or $\Omega_m \Omega_\ell = H_M$. □

Right triangles

An important property of right triangles in Euclidean geometry is given by Pythagoras' theorem. In spherical geometry we have the following analogue. See Figure 4.24.

Theorem 51. *Let ABC be a triangle on $\mathbf{S}^2$ with sides of length $a = d(B, C)$, $b = d(A, C)$, and $c = d(A, B)$. If AC is perpendicular to AB, then*

$$\cos a = \cos b \cos c.$$

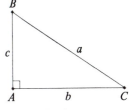

Figure 4.24 A right triangle, $\cos a = \cos b \cos c$.

Remark: Note that this is a special case of Theorem 38. However, a direct proof is instructive.

Proof: Let ξ be a pole of $\overleftrightarrow{AB}$. Then $\{\xi, A, \xi \times A\}$ is an orthonormal basis with respect to which we may write (after replacing ξ by $-\xi$ if necessary)

$$C = (\cos b)A + (\sin b)\xi.$$

Also,

$$B = (\cos c)A \pm (\sin c)(\xi \times A).$$

Hence,

$$\cos a = \cos d(B, C) = \cos \cos^{-1}\langle B, C \rangle$$
$$= \langle B, C \rangle = \cos b \cos c. \qquad \square$$

Theorem 52. *Let ℓ be any line. Let X be a point that is neither on ℓ nor a pole of ℓ. Let m be the line through X perpendicular to ℓ. Of the two points where ℓ intersects m, let F be the one closest to X. Then for all points $Y \neq \pm F$ on ℓ,*

$$d(X, F) < d(X, Y) < d(X, -F).$$

Proof: We apply Theorem 51 with $X = C$, $F = A$, and $Y = B$. Note that $b < \pi/2$, so that $\cos a$ and $\cos c$ have the same sign. If both are positive, we have

$$\cos a = \cos b \cos c < \cos b,$$

and, hence,

$$b < a < \pi - b.$$

If both are zero, the same inequality holds because

$$b < a = \pi/2 < \pi - b.$$

Finally, if both are negative, we get $b < \pi/2 < a$ and

115

$$\cos(\pi - a) = \cos b \cos(\pi - c) < \cos b,$$

so that $\pi - a > b$, and, hence, $b < a < \pi - b$, as required. ☐

Remark: This means that F is the point of ℓ closest to X, and $-F$ is the point farthest from X.

Definition. *F is called the* foot *of the perpendicular from X to ℓ. The number $d(X, F)$ is written $d(X, \ell)$ and is called the* distance *from X to ℓ.*

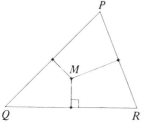

P

M

Q R

Figure 4.25 Concurrence of the perpendicular bisectors.

Concurrence theorems

Theorem 53. *The perpendicular bisectors of the three sides of a triangle are concurrent. See Figure 4.25.*

Proof: Let the triangle be $\triangle PQR$. Let M be a point where the perpendicular bisectors of sides PQ and QR intersect. Now $d(M, P) = d(M, Q)$ and $d(M, Q) = d(M, R)$. Thus, $d(M, P) = d(M, R)$, and M lies on the perpendicular bisector of side PR. ☐

Remark: The same theorem with the same proof is valid in $\mathbf{E}^2$.

Theorem 54. *Let P, Q, and R be noncollinear points of $\mathbf{S}^2$. Let $p = \overleftrightarrow{QR}$, $q = \overleftrightarrow{PR}$, and $r = \overleftrightarrow{PQ}$ be the three lines they determine. Let Ω_u be a reflection that interchanges p and q, and let Ω_v interchange q and r. Then there is a line w concurrent with u and v such that Ω_w interchanges p and r.*

Proof: Let M be a point of intersection of u and v, and let ℓ be the line through M perpendicular to q. Using the three reflections theorem, choose w so that

$$\Omega_v \Omega_\ell \Omega_u = \Omega_w.$$

Then

$$\Omega_w p = \Omega_v \Omega_\ell \Omega_u p = \Omega_v \Omega_\ell q = \Omega_v q = r,$$

as required. ☐

Corollary. *The lines containing the bisectors of the three angles of a triangle are concurrent.*

Proof: Because there are two reflections that interchange p and i, we need only check that w is the one containing the bisector of $\angle PQR$. This part of the proof requires some further calculation and will be left as an exercise (Exercise 40). Figures 4.26 and 4.27 illustrate the possibilities. □

Congruence theorems for triangles

Theorem 55 (SSS theorem). *Let $\triangle PQR$ and $\triangle P'Q'R'$ be such that $d(P, Q) = d(P', Q'), d(Q, R) = d(Q', R')$, and $d(P, R) = d(P', R')$. Then the two triangles are congruent.*

Theorem 56 (SAS theorem). *Let $\triangle PQR$ and $\triangle P'Q'R'$ be such that $d(P, Q) = d(P', Q'), d(Q, R) = d(Q', R')$, and $\angle PQR = \angle P'Q'R'$ (in radian measure). Then the two triangles are congruent.*

These two theorems are proved in a similar fashion to Theorems 1.40 and 1.41 by using the spherical versions of the tools used in the construction. Because sizes of angles determine lengths of sides in spherical geometry (Theorem 38), we get an additional congruence theorem.

Theorem 57 (AAA theorem). *Let $\triangle PQR$ and $\triangle P'Q'R'$ be such that $\angle PQR = \angle P'Q'R'$, $\angle PRQ = \angle P'R'Q'$, and $\angle QPR = \angle Q'P'R'$ (in radian measure). Then the two triangles are congruent.*

Corollary. *Two angles are congruent if and only if they have the same radian measure.*

Finite rotation groups

In plane Euclidean geometry we found a nice characterization of the finite groups that occur as symmetry groups of figures. All finite subgroups of $\mathscr{I}(\mathbf{E}^2)$ were shown to be cyclic or dihedral (Theorem 3.10).

The situation in spherical geometry is more complicated. In other words, figures can have a richer symmetry structure. The standard figures in $\mathbf{E}^2$ having the largest symmetry groups are the regular polygons. In spherical geometry we have, in addition to the regular polygons, the spherical versions of the Platonic solids – the tetrahedron, cube, octahedron, dodecahedron, and icosahedron.

We will restrict our attention to finite groups of rotations of $\mathbf{S}^2$. All other finite subgroups of $\mathbf{O(3)}$ are generated by adjoining a suitable reflection.

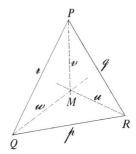

Figure 4.26 Concurrence of the angle bisectors.

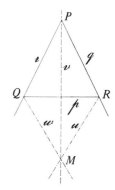

Figure 4.27 Concurrence of exterior angle bisectors with remote interior angle bisector.

Theorem 58. *Let G be a finite subgroup of* SO(3). *Then G falls into one of the categories listed in the following table.*

G	Order of G	Number of orbits	Number of poles	Orders of stabilizers		
cyclic	n	2	2	n	n	
dihedral	$2n$	3	$2n + 2$	2	2	n
tetrahedral	12	3	14	2	3	3
octahedral	24	3	26	2	3	4
icosahedral	60	3	62	2	3	5

Every finite rotation group is conjugate to one of the cyclic or dihedral groups or to one of the three specific groups listed. Our treatment is incomplete. We will show that any finite rotation group has data fitting the table, but we will not show uniqueness for these groups or, in fact, even define the groups explicitly. For more details you may consult Benson and Grove [4] or Yale [35].

We describe the cyclic and dihedral groups explicitly. With respect to an orthonormal basis $\{e_1, e_2, e_3\}$ let α be the rotation whose matrix is

$$\begin{bmatrix} \operatorname{rot} \dfrac{2\pi}{n} & 0 \\ 0 & 1 \end{bmatrix}.$$

Then α generates a cyclic group of order n consisting of rotations about e_3. Now let β be the half-turn about e_1. Then α and β generate a group satisfying the relations $\alpha^n = \beta^2 = I$, $\beta\alpha = \alpha^{-1}\beta$. It is easy to check that $\alpha^k\beta$ has the matrix

$$\begin{bmatrix} \cos \dfrac{2k\pi}{n} & \sin \dfrac{2k\pi}{n} & 0 \\ \sin \dfrac{2k\pi}{n} & -\cos \dfrac{2k\pi}{n} & 0 \\ 0 & 0 & -1 \end{bmatrix}$$

and so is a half-turn about the point $(\cos(k\pi/n), \sin(k\pi/n), 0)$.

Now let G be a finite subgroup of SO(3). If T is a nontrivial rotation about x, we call x a *pole* of T. If G has order n, there are $2(n-1)$ ordered pairs (T, x) consisting of a nontrivial rotation in G and one of its poles.

Lemma. *G permutes the set of poles.*

Proof: Let x be a pole of some rotation $T \in G$, and let R be an element of G. Then

$$Rx = RTx = (RTR^{-1})Rx.$$

Now RTR^{-1} is a rotation in G having Rx as a pole. Thus, R maps poles to poles. Similarly, $x = R(R^{-1}x)$, and $R^{-1}x$ is a pole of $R^{-1}TR$. We conclude that R is a permutation of the set of poles. $\qquad\square$

Proof (of Theorem 58): For each pole x let $v_x = \#\text{Orbit}(x)$, and $n_x = \#\text{Stab}(x)$. Then $n_x v_x = n = \#G$. Now counting the ordered pairs (T, x) gives

$$2(n - 1) = \sum_x (n_x - 1) = \sum_{i=1}^{k} v_i(n_i - 1),$$

where $\{x_1 \ldots x_k\}$ are representatives of the disjoint orbits that make up the set of poles. Writing $n_i = n_{x_i}$ and $v_i = v_{x_i}$, we get

$$2(n - 1) = \sum v_i n_i - \sum v_i = \sum_{i=1}^{k} (n - v_i).$$

Thus,

$$2 - \frac{2}{n} = \sum_i \left(1 - \frac{v_i}{n}\right) = \sum \left(1 - \frac{1}{n_i}\right). \qquad (4.6)$$

This formula will allow us to determine the number and size of the orbits. First, note that any fixed point of a nontrivial rotation T in G is also a fixed point of T^{-1}. Thus, we may conclude that $n_i \geq 2$ for all i. Clearly, the number k of orbits cannot be 1 because $2 - 2/n \geq 1$. On the other hand, we have

$$\frac{2}{n_i} \leq 1 \quad \text{and} \quad \frac{1}{2} \leq 1 - \frac{1}{n_i} < 1,$$

and, hence,

$$\frac{k}{2} \leq \sum_{i=1}^{k} \left(1 - \frac{1}{n_i}\right) < 2.$$

In particular, $k < 4$, and so the only possibilities for k are 2 and 3.
We first look at the case $k = 2$. Formula (4.6) reduces to

$$2 - \frac{2}{n} = 1 - \frac{1}{n_1} + 1 - \frac{1}{n_2},$$

$$\frac{2}{n} = \frac{1}{n_1} + \frac{1}{n_2},$$

$$2 = v_1 + v_2.$$

Thus, $v_1 = v_2 = 1$, $n_1 = n_2 = n$, and there are just two poles. G is a group of rotations around this pair of antipodal points.

The remaining possibility is $k = 3$. We may assume that $n_1 \leqslant n_2 \leqslant n_3$. We first note that $n_1 = 2$ because $n_1 \geqslant 3$ would give

$$\Sigma \left(1 - \frac{1}{n_i}\right) \geqslant \Sigma \left(1 - \frac{1}{3}\right) = 2 > 2 - \frac{1}{n},$$

violating (4.6). With this simplification (4.6) becomes

$$2 - \frac{2}{n} = \frac{1}{2} + \left(1 - \frac{1}{n_2}\right) + \left(1 - \frac{1}{n_3}\right);$$

that is,

$$\frac{2}{n} = \frac{1}{n_2} + \frac{1}{n_3} - \frac{1}{2}. \qquad (4.7)$$

In order to make the right side of (4.7) positive, we must have $n_2 \leqslant 3$. If $n_2 = 2$, then $2/n = 1/n_3$, and $v_3 = 2$. Also, $v_1 = v_2 = n/2$. On the other hand, $n_2 = 3$ yields

$$\frac{2}{n} = \frac{1}{3} + \frac{1}{n_3} - \frac{1}{2} = \frac{1}{n_3} - \frac{1}{6}, \qquad (4.8)$$

and so $n_3 < 6$. Thus, the remaining possibilities for n_3 are 3, 4, and 5. Unlike the previous cases, each possible value for n_3 uniquely determines the order n of the group G. Specifically, the possible combinations for (n_3, n) are $(3, 12)$, $(4, 24)$, and $(5, 60)$, as can be easily seen from (4.8). $\quad \square$

Finite groups of isometries of $\mathbf{S}^2$

Let G be a finite group of isometries of $\mathbf{S}^2$. Assume that G does not consist entirely of rotations. Choose an element β that is not a rotation. Then $G = G_0 \cup \beta G_0$, where G_0 is the set of rotations in G. Thus, the group G must have order $2n$, $4n$, 24, 48, or 120, depending on the structure of G_0. It is an interesting exercise to determine the group structures that can occur. Among the groups so obtained will be the symmetry groups of the regular polygons of $\mathbf{S}^2$, the "degenerate" regular polygons having all vertices collinear and the regular polyhedra (Platonic solids) discussed earlier in this chapter.

EXERCISES

1. Prove the properties of cross products stated in parts (i), (iii), and (iv) of the corollary to Theorem 1.

2. Prove Theorem 2.

3. Prove that lines $P + [v]$ and $Q + [w]$ intersect if and only if
$$\langle Q - P, v \times w \rangle = 0.$$

4. Assume that the lines of Exercise 3 do not intersect. Find the (shortest) distance between them.

5. Prove Theorem 5.

6. Given three points P, Q, and R of $\mathbf{S}^2$, what calculation can you perform to determine whether P, Q, and R are collinear? Apply it to the points

$$
\begin{bmatrix} \dfrac{1}{\sqrt{2}} \\[2mm] 0 \\[2mm] \dfrac{1}{\sqrt{2}} \end{bmatrix}, \quad
\begin{bmatrix} \dfrac{1}{\sqrt{5}} \\[2mm] \dfrac{2}{\sqrt{5}} \\[2mm] 0 \end{bmatrix}, \quad \text{and} \quad
\begin{bmatrix} 0 \\[2mm] \dfrac{-2}{\sqrt{5}} \\[2mm] \dfrac{1}{\sqrt{5}} \end{bmatrix}.
$$

7. Let A and B be distinct points of $\mathbf{S}^2$. Show that
$$\{X \in \mathbf{S}^2 | d(X, A) = d(X, B)\}$$
is a line, and find an expression for its pole.

8. Verify statements (i)–(iii) of Theorem 8.

9. Prove Theorem 10.

10. Prove Theorem 11. In particular, show that
   i. The poles of m are $\pm(\xi \times P)/|\xi \times P|$, where ξ is a pole of ℓ.
   ii. The points of intersection are
   $$\pm\frac{P - \langle P, \xi \rangle \xi}{(1 - \langle P, \xi \rangle^2)^{1/2}}.$$
   iii. The distances from P to ℓ are
   $$\cos^{-1}(\pm(1 - \langle P, \xi \rangle^2)^{1/2}).$$

11. Prove Theorem 12, part (i) and Theorem 13.

12. Verify that Theorems 15 and 16 can be proved with the same calculations as were used in the Euclidean case.

13. Prove Theorem 21.

14. Let P and Q be distinct nonantipodal points. Under what circumstances will the group generated by $\{H_P, H_Q\}$ be finite? (*Note*: A half-turn on $\mathbf{S}^2$ is again a product of reflections in two perpendicular lines.)

15. Prove Theorem 27 concerning fixed points of isometries.

16. Prove Theorems 28 and 29 concerning fixed lines of isometries.

17. If an isometry of $\mathbf{S}^2$ leaves P fixed and takes Q to $-Q$, show that $\langle P, Q \rangle = 0$.

18. Let $P = (1, 1, 0)$ and $Q = (3, 2, 1)$ be points of $\mathbf{E}^3$. Let $X = P/|P|$, $Y = Q/|Q|$.

    i. Find an orthonormal basis with X as one element and the pole of $\overleftrightarrow{XY}$ as another.

    ii. Compute the matrices of H_X and H_Y (as isometries of $\mathbf{S}^2$) with respect to this basis.

19. Classify the isometries α of $\mathbf{S}^2$ satisfying $\alpha^4 = I$. (You may use the fact that isometries, motions, and orthogonal transformations are essentially the same thing.)

20. Find all isometries α of $\mathbf{S}^2$ such that $\alpha^2 = I$, but $\alpha \neq I$. Such an isometry is said to be an *involution*. If α and β are involutions, is $\alpha\beta$ an involution?

21. Let P be a point of $\mathbf{S}^2$, and ℓ a line of $\mathbf{S}^2$. Show that

$$(\Omega_\ell H_P)^2 = I$$

    if and only if P is a pole of ℓ or $P \in \ell$.

22. If $a \perp \mathcal{b}$ and $\mathcal{b} \perp c$, what is $\Omega_a \Omega_\mathcal{b} \Omega_c$?

23. Verify the following formula for a half-turn:

$$H_P X = 2\langle X, P \rangle P - X.$$

24. Let ℓ be a line of $\mathbf{S}^2$ with pole P. Show that $\{I, H_P, \Omega_\ell, E\}$ is a group, and give its multiplication table. (E is the antipodal map.)

25. Without using Euler's theorem, show that the product of two rotations is a rotation.

26. Let γ be a glide reflection. Prove that γE has exactly two (antipodal) fixed points unless γ is a reflection or $\gamma = E$.

27. Let ℓ be a line. Find $\mathscr{S}(\ell)$.

28. Let $\mathscr{F} = \{P, Q, R\}$ be a figure consisting of three mutually perpendicular points. Find $\mathscr{S}(\mathscr{F})$.

29. Under what circumstances will a reflection and a half-turn commute?

30. Prove or disprove the formula

$$H_P H_Q H_R = H_R H_Q H_P.$$

31. Let P be a point of $\mathbf{S}^2$. Show that the stabilizer of P in $\mathbf{O}(3)$ consists of the rotations about P and the reflections in lines through P.

32. Let ℓ, m, and n be mutually perpendicular lines, and let L, M, and N be respective intersection points $m \cap n$, $n \cap \ell$, and $\ell \cap m$. Show that $\{\Omega_\ell, \Omega_m, \Omega_n, H_L, H_M, H_N, E, I\}$ is a group and that $\{H_L, H_M, H_N; I\}$ is a subgroup.

33. Fill in the missing argument in the proof of Theorem 35.

34. Let P, Q, and R be distinct points of S^2. Assume neither Q nor R are antipodal to P. Prove that $\overrightarrow{PQ} = \overrightarrow{PR}$ if and only if P, Q, and R are collinear, Q lies on PR, or R lies on PQ. (PR and PQ are taken to be minor segments.)

35. Verify that $\{X | \langle X, \xi \rangle \geq 0\}$ is a half-plane. Does the crossbar theorem hold in S^2?

36. Although a given angle can be represented in many ways, the definition of its radian measure is independent of the representation. Prove this.

37. Complete the proof of Theorem 38.

38. Adapt the Euclidean material on rectilinear figures (Chapter 2, Theorems 24, 25, and their corollaries) to spherical geometry. Verify that spherical triangles have the same symmetry properties as Euclidean triangles.

39. Verify the remark following Theorem 46.

40. Show that the line ω in the corollary to Theorem 54 contains the bisector of $\sphericalangle PQR$.

41. Prove that the product of reflections in the perpendicular bisectors of the sides of a triangle is a reflection whose axis passes through a vertex.

5

The projective plane $\mathbf{P}^2$

Introduction

We now come to a geometric structure that is more abstract than the previous two we have dealt with. The geometry of the projective plane will resemble that of the sphere in many respects. However, we regain the Euclidean phenomenon that two lines can intersect only once. The projective plane will also be a foundation for our study of hyperbolic geometry in Chapter 7.

Although many of the properties of the projective plane are familiar, one that will appear strange is that of nonorientability. In $\mathbf{P}^2$ every reflection may be regarded as a rotation. This has the intuitive consequence that an outline of a left hand can be moved continuously to coincide with its mirror image, the outline of a right hand.

The abstraction is involved in the fact that every point of $\mathbf{P}^2$ is a *pair of points* of $\mathbf{S}^2$. Two antipodal points of $\mathbf{S}^2$ are considered to be the same point of $\mathbf{P}^2$.

Definition. *The projective plane* $\mathbf{P}^2$ *is the set of all pairs* $\{x, -x\}$ *of antipodal points of* $\mathbf{S}^2$.

Remark: Two alternative definitions of $\mathbf{P}^2$, equivalent to the preceding one are
i. The set of all lines through the origin in $\mathbf{E}^3$.
ii. The set of all equivalence classes of ordered triples (x_1, x_2, x_3) of numbers (i.e., vectors in $\mathbf{E}^3$) not all zero, where two vectors are equivalent if they are proportional.

Let $\pi: \mathbf{S}^2 \to \mathbf{P}^2$ be the mapping that sends each x to $\{x, -x\}$. Then π is a two-to-one map of $\mathbf{S}^2$ onto $\mathbf{P}^2$.

A *line* of $\mathbf{P}^2$ is a set of the form $\pi\ell$, where ℓ is a line of $\mathbf{S}^2$. If ξ is a pole of ℓ, then $\pi\xi$ is called the *pole* of $\pi\ell$. Clearly, πx lies on $\pi\ell$ if and only if $\langle \xi, x \rangle = 0$. Two points are perpendicular if their representatives on $\mathbf{S}^2$ are perpendicular. Two lines are perpendicular if their poles are perpendicular.

Incidence properties of $\mathbf{P}^2$

Theorem 1.
i. *Two lines of $\mathbf{P}^2$ have exactly one point of intersection.*
ii. *Two points of $\mathbf{P}^2$ lie on exactly one line.*

Proof:
i. Let $\pi\xi$ and $\pi\eta$ be poles of lines of $\mathbf{P}^2$. Because $\pi\xi \neq \pi\eta$, ξ and η are not antipodal. Thus, $\xi \times \eta$ and $-\xi \times \eta$ determine the two points of intersection of the corresponding lines of $\mathbf{S}^2$ (Theorem 4.7). But $\pi(\xi \times \eta)$ and $\pi(-\xi \times \eta)$ are the same point of $\mathbf{P}^2$.
ii. Again, let πX and πY be points of $\mathbf{P}^2$. Then X and Y are not antipodal, so they lie on a unique line ℓ of $\mathbf{S}^2$ (Theorem 4.6). Thus, πX and πY lie on $\pi\ell$. $\qquad\square$

Homogeneous coordinates

Let $\{e_1, e_2, e_3\}$ be a basis of $\mathbf{R}^3$. Then every vector $x \in \mathbf{R}^3$ determines a unique triple (x_1, x_2, x_3) of real numbers according to the equation

$$x = x_1 e_1 + x_2 e_2 + x_3 e_3.$$

If πx is a point of $\mathbf{P}^2$, λ is any nonzero real number, and

$$\lambda x = u_1 e_1 + u_2 e_2 + u_3 e_3,$$

then (u_1, u_2, u_3) is called a *homogeneous coordinate vector* of πx. We say that u_1, u_2, u_3 are *homogeneous coordinates* of πx.

Let $\xi = (\xi_1, \xi_2, \xi_3)$ and $x = (x_1, x_2, x_3)$. Then $\langle \xi, x \rangle = 0$ becomes the equation of the line with pole $\pi\xi$. Homogeneous coordinates are often a useful computational device. Their usefulness is primarily due to the following result.

Theorem 2. *Let P, Q, R, and S be four points of $\mathbf{P}^2$, no three of which are collinear. Then there is a basis of $\mathbf{R}^3$ with respect to which the four points have coordinates $(1, 0, 0)$, $(0, 1, 0)$, $(0, 0, 1)$, and $(1, 1, 1)$.*

Proof: Let v_1, v_2, and v_3 be any vectors in $\mathbf{R}^3$ that are representatives of P, Q, and R, respectively. Because P, Q, and R are not collinear, these three vectors are linearly independent. Let v_4 be any representative of S. Now there must exist real numbers k_1, k_2, k_3, none of which is zero, such that

$$v_4 = k_1v_1 + k_2v_2 + k_3v_3.$$

Put $e_1 = k_1v_1$, $e_2 = k_2v_2$, and $e_3 = k_3v_3$. Then $\{e_1, e_2, e_3\}$ is the required basis. $\qquad\square$

Theorem 3. *Let x and y be homogeneous coordinate vectors of two points of $\mathbf{P}^2$. Then $\lambda x + \mu y$ (λ, μ real) is a typical point on the line they determine.*

Two famous theorems

Having introduced the incidence structure of $\mathbf{P}^2$ and having defined the notion of homogeneous coordinates, we turn to two fundamental classical theorems in projective geometry: Desargues' theorem and Pappus' theorem. The elegance of the statements testifies to the unifying power of projective geometry. Analogous results in $\mathbf{E}^2$ would have to make allowances for many special cases. The elegance of the proofs (which follow Coxeter [7]) testifies to the power of the method of homogeneous coordinates. In this section the word "triangle" denotes a set of three noncollinear points. We have not yet defined segments in $\mathbf{P}^2$, so our old notion of triangle does not apply.

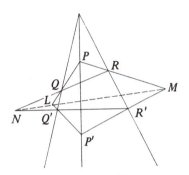

Figure 5.1 Desargues' theorem.

Theorem 4 (Desargues' theorem). *Let PQR and $P'Q'R'$ be triangles in $\mathbf{P}^2$. Suppose $\overleftrightarrow{PP'}$, $\overleftrightarrow{QQ'}$, and $\overleftrightarrow{RR'}$ are concurrent. Then $\overleftrightarrow{PQ} \cap \overleftrightarrow{P'Q'}$, $\overleftrightarrow{QR} \cap \overleftrightarrow{Q'R'}$, and $\overleftrightarrow{PR} \cap \overleftrightarrow{P'R'}$ are collinear. (See Figure 5.1.)*

Proof: We may choose a basis for $\mathbf{R}^3$ such that in the associated homogeneous coordinate system $P = (1, 0, 0)$, $Q = (0, 1, 0)$, $R = (0, 0, 1)$, and $X = (1, 1, 1)$, where X is the given point of concurrence. If X were collinear with any two of these points, then two sides (such as $\overleftrightarrow{PQ}$ and $\overleftrightarrow{P'Q'}$) would coincide, leaving the conclusion meaningless. Thus, we may assume that no three of P, Q, R, and X are collinear. Now P' may be given coordinates $(p, 1, 1)$ because

$$\lambda(1, 0, 0) + \mu(1, 1, 1) = (\lambda + \mu, \mu, \mu),$$

which is equivalent to

$$\left(1 + \frac{\lambda}{\mu}, 1, 1\right).$$

Similarly, $Q' = (1, q, 1)$ and $R' = (1, 1, r)$. Now the equation of $\overleftrightarrow{PQ}$ is $x_3 = 0$, and that of $\overleftrightarrow{P'Q'}$ is

$$(1 - q)x_1 + (1 - p)x_2 + (pq - 1)x_3 = 0.$$

These lines intersect in $L = (p - 1, 1 - q, 0)$. Similarly, the other two points of intersection are $M = (1 - p, 0, r - 1)$ and $N = (0, q - 1, 1 - r)$. The three points L, M, and N are collinear because the sum of the three coordinate vectors is zero. □

Theorem 5 (Pappus' theorem). *Let $A_1B_1C_1$ and $A_2B_2C_2$ be collinear triples of points. Then the points $\overleftrightarrow{A_1B_2} \cap \overleftrightarrow{A_2B_1} = C_3$, $\overleftrightarrow{B_2C_1} \cap \overleftrightarrow{B_1C_2} = A_3$, and $\overleftrightarrow{A_1C_2} \cap \overleftrightarrow{A_2C_1} = B_3$ are collinear. (See Figure 5.2.)*

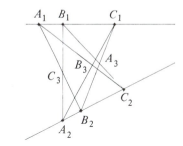

Figure 5.2 Pappus' theorem.

Proof: Assign homogeneous coordinates as follows:

$$A_1 = (1, 0, 0), \quad A_2 = (0, 1, 0), \quad A_3 = (0, 0, 1),$$
$$C_1 = (1, 1, 1), \quad B_1 = (p, 1, 1), \quad B_3 = (1, q, 1),$$
$$B_2 = (1, 1, r).$$

Then

$$C_2 = \overleftrightarrow{B_1A_3} \cap \overleftrightarrow{B_3A_1} = (pq, q, 1),$$
$$C_3 = \overleftrightarrow{A_1B_2} \cap \overleftrightarrow{A_2B_1} = (pr, 1, r).$$

Because A_2, B_2, and C_2 are collinear, we must have

$$C_2 = (0, \lambda, 0) + (1, 1, r) = (1, \lambda + 1, r).$$

On the other hand,

$$C_2 = (pq, q, 1) = (pqr, qr, r).$$

Thus, we must have $pqr = 1$.

Now $\overleftrightarrow{A_3B_3}$ consists of points of the form $(0, 0, \lambda) + (1, q, 1) = (1, q, 1 + \lambda)$. Because $C_3 = (pqr, q, rq) = (1, q, rq)$, it must be on this line. □

Applications to $\mathbf{E}^2$

One of the reasons for the invention of $\mathbf{P}^2$ was to simplify the incidence geometry of $\mathbf{E}^2$. To illustrate this, consider the following picture in $\mathbf{E}^3$. We regard the plane $x_3 = 1$ consisting of all points in $\mathbf{E}^3$ of the form $(x_1, x_2, 1)$ as a model of $\mathbf{E}^2$. Every line through the origin of $\mathbf{E}^3$ that is not parallel to $\mathbf{E}^2$ meets $\mathbf{E}^2$ in a unique point. If (x_1, x_2, x_3) are homogeneous coordinates for such a point of $\mathbf{P}^2$, then

$$\left(\frac{x_1}{x_3}, \frac{x_2}{x_3}, 1 \right) \tag{5.1}$$

The projective plane $\mathbf{P}^2$

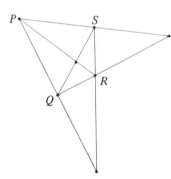

Figure 5.3   Quadrangle: Case 1.

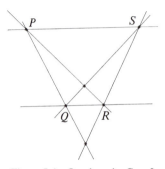

Figure 5.4   Quadrangle: Case 2.

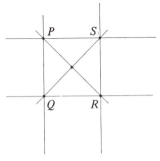

Figure 5.5   Quadrangle: Case 3.

128

is the corresponding point of $\mathbf{E}^2$. Conversely, each point of $\mathbf{E}^2$ determines a unique point of $\mathbf{P}^2$.

Every line of $\mathbf{E}^2$ determines a unique plane through the origin in $\mathbf{E}^3$ and, hence, a unique line of $\mathbf{P}^2$. Every line of $\mathbf{P}^2$ determines a unique plane through the origin in $\mathbf{E}^3$ and, hence (with one exception), a unique line of $\mathbf{E}^2$. The exception is the plane through the origin parallel to $\mathbf{E}^2$.

Let $T: \mathbf{E}^2 \to \mathbf{P}^2$ be the map we have been discussing.

Theorem 6.

i.   *Denote by ℓ_∞ the exceptional line of $\mathbf{P}^2$. Then T maps $\mathbf{E}^2$ bijectively to $\mathbf{P}^2 - \ell_\infty$.*

ii. *Let P and Q be points of $\mathbf{E}^2$. Then TP and TQ determine a line ℓ' of $\mathbf{P}^2$, and T maps $\ell = \overleftrightarrow{PQ}$ bijectively to $\ell' - \ell_\infty$.*

iii. *Let ℓ and m be lines of $\mathbf{E}^2$. If $\ell \cap m = P$, then $\ell' \cap m' = TP$. If $\ell \parallel m$, then $\ell' \cap m'$ lies on ℓ_∞.*

Remark: What this theorem says is that $\mathbf{P}^2$ contains a subset that has the same incidence structure as $\mathbf{E}^2$. Two lines will be parallel on $\mathbf{E}^2$ if and only if they correspond to lines meeting on ℓ_∞.

Example: A quadrangle $PQRS$ in $\mathbf{P}^2$ consists of four points, no three collinear, and the six lines drawn through pairs of vertices. The three points $\overleftrightarrow{PQ} \cap \overleftrightarrow{RS}$, $\overleftrightarrow{PR} \cap \overleftrightarrow{QS}$, and $\overleftrightarrow{PS} \cap \overleftrightarrow{QR}$ are called *diagonal* points of the quadrangle. Now the corresponding figure in $\mathbf{E}^2$ can take on many forms, depending on where ℓ_∞ intersects the figure. We list the possibilities. They are illustrated in Figures 5.3–5.7.

1. ℓ_∞ contains no vertex (P, Q, R, or S) and no diagonal point. In this case we have an ordinary Euclidean quadrangle.
2. ℓ_∞ contains no vertex but one diagonal point. In this case two sides of the quadrangle are parallel; the other two are not.
3. ℓ_∞ contains no vertex but two diagonal points. In this case we have a parallelogram.
4. ℓ_∞ contains one vertex and no diagonal points. Here we have three ordinary points Q, R, and S, the lines $\overleftrightarrow{QR}$ and $\overleftrightarrow{RS}$, together with parallel lines through Q and S, respectively.
5. ℓ_∞ contains two vertices P and Q. In this case one diagonal point is forced to be on ℓ_∞.

If we start with the general case (1) and gradually turn one of the lines, say $\overleftrightarrow{PS}$, while leaving the others fixed, the point of intersection $\overleftrightarrow{PS} \cap \overleftrightarrow{QR}$ gets farther and farther away in $\mathbf{E}^2$. On $\mathbf{P}^2$ the corresponding point is getting closer to ℓ_∞. This is why ℓ_∞ is sometimes called "the line at ∞." This also accounts for the statement "parallel lines meet at ∞."

The projective version of Desargues' theorem has many interpretations in $\mathbf{E}^2$, depending on where the various lines cut ℓ_∞. For instance, if X is on ℓ_∞, the theorem would read as follows.

Theorem 7. *Let PQR and $P'Q'R'$ be triangles in $\mathbf{E}^2$. Suppose that $\overleftrightarrow{PP'}$, $\overleftrightarrow{QQ'}$, and $\overleftrightarrow{RR'}$ are parallel. Then*

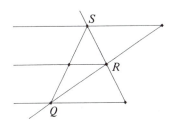

i. *If $\overleftrightarrow{PQ} \parallel \overleftrightarrow{P'Q'}$ and $\overleftrightarrow{QR} \parallel \overleftrightarrow{Q'R'}$, then $\overleftrightarrow{PR} \parallel \overleftrightarrow{P'R'}$ (Figure 5.8).*
ii. *If $\overleftrightarrow{PQ} \parallel \overleftrightarrow{P'Q'}$ but $\overleftrightarrow{QR} \cap \overleftrightarrow{Q'R'} = N$, then $\overleftrightarrow{PR}$ and $\overleftrightarrow{P'R'}$ meet (say in M), and $\overleftrightarrow{MN}$ is parallel to $\overleftrightarrow{PQ}$ (Figure 5.9).*
iii. *If $\overleftrightarrow{PQ} \cap \overleftrightarrow{P'Q'} = L$, $\overleftrightarrow{QR} \cap \overleftrightarrow{Q'R'} = M$ and $\overleftrightarrow{PR} \cap \overleftrightarrow{P'R'} = N$, then L, M, and N are collinear (Figure 5.10).*

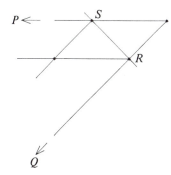

Figure 5.6   Quadrangle: Case 4.

Observe that the three cases correspond to the following in $\mathbf{P}^2$.

i. ℓ_∞ contains all three of L, M, and N.
ii. ℓ_∞ contains one of L, M, or N.
iii. ℓ_∞ contains none of L, M, or N.

If X is not on ℓ_∞ in Desargues' theorem, it would read as follows.

Theorem 8. *Let PQR and $P'Q'R'$ be triangles in $\mathbf{E}^2$. Suppose $\overleftrightarrow{PP'}$, $\overleftrightarrow{QQ'}$, and $\overleftrightarrow{RR'}$ meet in X. Then the conclusions of Theorem 7 hold.*

If we take ℓ_∞ to be the line $PP'X$ in Desargues' theorem, we get the following:

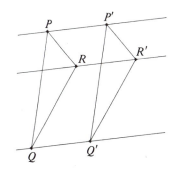

Figure 5.7   Quadrangle: Case 5.

Theorem 9. *Let $QRR'Q'$ be a trapezoid ($\overleftrightarrow{QQ'} \parallel \overleftrightarrow{RR'}$). Let ℓ and m be parallel lines through Q and R. Let ℓ' and m' be parallel lines through Q' and R'. Let $X = \ell \cap \ell'$, $Y = m \cap m'$, and $Z = \overleftrightarrow{QR} \cap \overleftrightarrow{Q'R'}$. Then X, Y, and Z are collinear.*

Theorem 10. *Let $QRR'Q'$ be a parallelogram in $\mathbf{E}^2$ ($\overleftrightarrow{QQ'} \parallel \overleftrightarrow{RR'}$ and $\overleftrightarrow{Q'R'} \parallel \overleftrightarrow{QR}$). Let ℓ and m be parallel lines through Q and R. Let ℓ' and m' be parallel lines through Q' and R'. If $X = \ell \cap \ell'$ and $Y = m \cap m'$, then $\overleftrightarrow{XY}$ is parallel to $\overleftrightarrow{QR}$.*

The projective group

Figure 5.8   Affine consequences of Desargues' theorem: Case 1.

Let **PGL(2)** be the group of collineations of $\mathbf{P}^2$. (Use the same definition as for $\mathbf{E}^2$.) Each invertible linear map $A: \mathbf{R}^3 \to \mathbf{R}^3$ determines a unique collineation $\tilde{A}$ in **PGL(2)** according to the definition

129

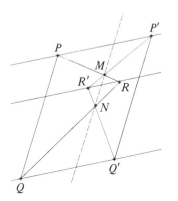

Figure 5.9 Affine consequences of Desargues' theorem: Case 2.

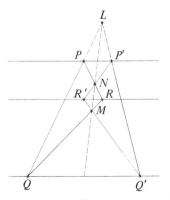

Figure 5.10 Affine consequences of Desargues' theorem: Case 3.

$$\tilde{A}\pi x = \pi A x. \tag{5.2}$$

The mapping $A \to \tilde{A}$ is a homomorphism of $\mathbf{GL(3)} \to \mathbf{PGL(2)}$ whose kernel is

$$K = \{kI \mid k \neq 0 \in \mathbf{R}\}.$$

It is a fact that this map is surjective (Exercise 17), so that

$$\mathbf{GL(3)}/K \cong \mathbf{PGL(2)}.$$

This fact is equivalent to the characterization of affine transformations in Theorem 2.2, whose proof is given in Appendix E.

Now if $A = kI$, then $\det A = k^3$. Because $k^3 = 1$ if and only if $k = 1$, we see that

1. Every member of $\mathbf{GL(3)}$ is equivalent to (is a multiple of) some member of $\mathbf{SL(3)}$. In particular, if $k = (\det A)^{-1/3}$, we see that $\det(kA) = 1$, so that $kA \in \mathbf{SL(3)}$.
2. $\mathbf{SL(3)} \cap K = \{I\}$.

Thus, the homomorphism restricted to $\mathbf{SL(3)}$ is an isomorphism, and

$$\mathbf{SL(3)} \cong \mathbf{PGL(2)}.$$

The subgroup of $\mathbf{PGL(2)}$ that fixes ℓ_∞ may be identified with $\mathbf{AF(2)}$. In fact, it is the image of $\mathbf{AF(2)}$ under the composition of the usual mappings:

$$\mathbf{AF(2)} \to \mathbf{GL(3)} \to \mathbf{PGL(2)} \tag{5.3}$$

It is easy to check that this composite mapping is injective.

An element of $\mathbf{PGL(2)}$ is called a *projective collineation* or projective transformation.

The fundamental theorem of projective geometry

Theorem 11. *Let PQRS and P'Q'R'S' be quadrangles. Then there is a unique $T \in \mathbf{PGL(2)}$ such that $TP = P'$, $TQ = Q'$, $TR = R'$, and $TS = S'$.*

Proof: Choose homogeneous coordinates of $(1, 0, 0)$, $(0, 1, 0)$, $(0, 0, 1)$, and $(1, 1, 1)$ for P, Q, R, and S, respectively. Then let A be a matrix whose columns are coordinate vectors for P', Q', and R', respectively. Call them x, y, and z. Now

$$\begin{bmatrix} x_1 & y_1 & z_1 \\ x_2 & y_2 & z_2 \\ x_3 & y_3 & z_3 \end{bmatrix} \begin{bmatrix} 1 & 0 & 0 \\ 0 & 1 & 0 \\ 0 & 0 & 1 \end{bmatrix} = \begin{bmatrix} x_1 & y_1 & z_1 \\ x_2 & y_2 & z_2 \\ x_3 & y_3 & z_3 \end{bmatrix}.$$

However,

$$\begin{bmatrix} x_1 & y_1 & z_1 \\ x_2 & y_2 & z_2 \\ x_3 & y_3 & z_3 \end{bmatrix}\begin{bmatrix} 1 \\ 1 \\ 1 \end{bmatrix} = \begin{bmatrix} x_1 + y_1 + z_1 \\ x_2 + y_2 + z_2 \\ x_3 + y_3 + z_3 \end{bmatrix},$$

and, for any λ, μ, ν,

$$\begin{bmatrix} \lambda x_1 & \mu y_1 & \nu z_1 \\ \lambda x_2 & \mu y_2 & \nu z_2 \\ \lambda x_3 & \mu y_3 & \nu z_3 \end{bmatrix}\begin{bmatrix} 1 \\ 1 \\ 1 \end{bmatrix} = \begin{bmatrix} \lambda x_1 + \mu y_1 + \nu z_1 \\ \lambda x_2 + \mu y_2 + \nu z_2 \\ \lambda x_3 + \mu y_3 + \nu z_3 \end{bmatrix}.$$

Choose λ, μ, and ν so that

$$w = \lambda x + \mu y + \nu z$$

is a coordinate vector for S'. The projective collineation whose matrix with respect to P, Q, R, and S is

$$\begin{bmatrix} \lambda x_1 & \mu y_1 & \nu z_1 \\ \lambda x_2 & \mu y_2 & \nu z_2 \\ \lambda x_3 & \mu y_3 & \nu z_3 \end{bmatrix}$$

is the required transformation T. Uniqueness will be proved in Exercise 16.

□

Corollary. *Let $\{P, Q, R\}$ and $\{P', Q', R'\}$ be two noncollinear triples of points. Let ℓ be a line not containing any of these points. Then there is a unique projective collineation T such that $TP = P'$, $TQ = Q'$, $TR = R'$, and $T\ell = \ell$.*

Proof: Let $\overleftrightarrow{PQ}$ and $\overleftrightarrow{P'Q'}$ meet ℓ in A and A', respectively. Let $\overleftrightarrow{PR}$ and $\overleftrightarrow{P'R'}$ meet ℓ in B and B', respectively. Then $RQAB$ and $R'Q'A'B'$ are quadrangles to which the fundamental theorem may be applied. The unique T so determined leaves ℓ fixed. Furthermore, because $P = \overleftrightarrow{RB} \cap \overleftrightarrow{QA}$, TP must be $\overleftrightarrow{R'B'} \cap \overleftrightarrow{Q'A'} = P'$.

Conversely, any projective collineation satisfying the stated conditions must take A to A' and B to B' and so must coincide with T. □

Remark: When a choice of ℓ_∞ has been made, this corollary with $\ell = \ell_\infty$ is just the fundamental theorem of affine geometry (Theorem 2.8).

A survey of projective collineations

In this section we will outline some of the facts about projective collineations. This material would occupy a whole chapter if done in detail.

Because our main emphasis in this book is on metric geometry, we will present the results with a minimum of discussion. All the necessary background for proving the theorems as exercises has already been developed.

Theorem 12. *Every projective collineation has at least one fixed point and one fixed line.*

In view of Theorem 12 it is useful to choose a point P and a line ℓ and examine the group of all projective collineations, leaving them fixed. We choose a homogeneous coordinate system in which P and ℓ have simple representations. If P lies on ℓ, let $P = (1, 0, 0)$. If P does not lie on ℓ, take $P = (0, 0, 1)$. In either case we can arrange that ℓ has the equation $x_3 = 0$. The next few theorems assume a homogeneous coordinate system satisfying these conditions.

Theorem 13. *If P does not lie on ℓ, the group of projective collineations leaving P and ℓ fixed is isomorphic to $\mathbf{GL(2)}$. Each such collineation can be uniquely written in the form*

$$\begin{bmatrix} a & b & 0 \\ c & d & 0 \\ 0 & 0 & 1 \end{bmatrix}, \quad ad - bc \neq 0.$$

Theorem 14. *Suppose that P lies on ℓ. Then every projective collineation leaving P and ℓ fixed is uniquely represented by a matrix of the form*

$$\begin{bmatrix} a & b & p \\ 0 & c & q \\ 0 & 0 & 1 \end{bmatrix}, \quad ac \neq 0.$$

Conversely, each such matrix determines a projective collineation leaving P and ℓ fixed.

Taking $\ell = \ell_\infty$ allows us to regard this group as the group of affine transformations leaving fixed one particular pencil of parallels, namely, the lines parallel to the x_1-axis. In fact, if two points (λ, μ) and $(\bar{\lambda}, \mu)$ in $\mathbf{E}^2$ have the same x_2-coordinate, then

$$\begin{bmatrix} a & b & p \\ 0 & c & q \\ 0 & 0 & 1 \end{bmatrix} \begin{bmatrix} \lambda & \bar{\lambda} \\ \mu & \mu \\ 1 & 1 \end{bmatrix} = \begin{bmatrix} a\lambda + b\mu + 1 & a\bar{\lambda} + b\mu + 1 \\ c\mu + q & c\mu + q \\ 1 & 1 \end{bmatrix},$$

so that their images also have the same x_2-coordinate. In affine terms this transformation is a central dilatation (centered at the origin), followed by a shear, and then a translation.

Of course, the transformation of Theorem 14 may have fixed points other than P and/or fixed lines other than ℓ. In fact,

Theorem 15. *The transformation of Theorem 14 has exactly one fixed point and one fixed line if and only if $a = c = 1$ and $bq \neq 0$.*

Theorem 16. *A projective collineation with two fixed points may be written in the form*

$$\begin{bmatrix} a & 0 & p \\ 0 & c & q \\ 0 & 0 & 1 \end{bmatrix}.$$

Such a collineation has at least two fixed lines.

Corollary. *A projective collineation with two fixed points may be written in the form*

$$\begin{bmatrix} a & 0 & 0 \\ 0 & c & q \\ 0 & 0 & 1 \end{bmatrix}.$$

In this representation $(1, 0, 0)$ and $(0, 1, 0)$ are fixed points. The lines $x_1 = 0$ and $x_3 = 0$ are fixed lines.

Theorem 17. *If a projective collineation has three collinear fixed points, it may be written*

$$\begin{bmatrix} a & 0 & 0 \\ 0 & a & q \\ 0 & 0 & 1 \end{bmatrix}, \quad a \neq 0.$$

Every point on the line $x_1 = 0$ is fixed. In addition, the line $x_3 = 0$ is a fixed line.

The transformations of Theorem 17 are called perspective collineations.

A *perspective collineation* with axis ℓ and center P is a projective collineation that leaves fixed every point on ℓ and every line through P. We may regard the identity as the trivial perspective collineation. All other perspective collineations have a unique axis and a unique center. A nontrivial perspective collineation is called an *elation* if its axis and center are incident; otherwise, it is called a *homology*.

Theorem 18. *When a perspective collineation is represented as in Theorem 17, it is*

i. *an elation if $a = 1$ and $q \neq 0$;*

ii. *the identity if $a = 1$ and $q = 0$;*
iii. *a homology if $a \neq 1$.*

Remark. If ℓ_∞ is taken to be the axis of a perspective collineation, elations become translations and homologies become central dilatations. On the other hand, if ℓ_∞ is one of the other fixed lines, elations become shears and homologies become stretches along one direction (see Theorem 2.20, case (iv)) possibly composed with an affine reflection. The special homology giving rise to an affine reflection is called a *harmonic homology.*

The term "perspective collineation" is explained by the following theorem.

Theorem 19. *Suppose that a nontrivial perspective collineation with center P takes X to X'. Then P, X, and X' are collinear.*

Theorem 20. *Let P be a point and ℓ a line. Let X and X' be points collinear with P. Assume that X and X' are not on ℓ and not equal to P. Then there is a unique perspective collineation with center P and axis ℓ that takes X to X'.*

Polarities

Let b be a real-valued, symmetric, nondegenerate, bilinear function on $\mathbf{E}^3$. If $\{e_1, e_2, e_3\}$ is a basis for $\mathbf{R}^3$. we have

$$b(x, y) = \sum_{i,j=1}^{3} x_i y_j b(e_i, e_j)$$

$$= \sum_{i,j=1}^{3} b_{ij} x_i y_j = x'By = \langle x, By \rangle, \tag{5.4}$$

where $B = [b_{ij}] = [b(e_i, e_j)]$.
Each such b determines a relation

$$\bar{b} \subset \mathbf{P}^2 \times \mathbf{P}^2$$

consisting of those pairs $(\pi x, \pi y)$ such that $b(x, y) = 0$.
The relation $\bar{b}$ is called a *polarity*. If $b(x, y) = 0$, we say that πx and πy are *conjugate*. For a given y the set

$$\{\pi x | b(x, y) = 0\}$$

is a line called the *polar line* of πy. We call πy the *pole* of the line with respect to b.

Some polarities have self-conjugate points. The set of self-conjugate points is called a *conic* determined by the polarity. For example, if

$$B = \begin{bmatrix} 1 & 0 & 0 \\ 0 & 1 & 0 \\ 0 & 0 & -1 \end{bmatrix},$$

then the conic determined by $\bar{b}$ is

$$\{\pi x | x'Bx = 0\} = \{\pi x | x_1^2 + x_2^2 - x_3^2 = 0\}.$$

This conic in $\mathbf{P}^2$ corresponds by formula (5.1) to the unit circle $x_1^2 + x_2^2 = 1$ in $\mathbf{E}^2$.

Similarly,

$$\begin{bmatrix} 2 & 0 & 0 \\ 0 & 3 & 0 \\ 0 & 0 & -4 \end{bmatrix} \quad \text{gives the ellipse} \quad 2x_1^2 + 3x_2^2 = 4,$$

$$\begin{bmatrix} -2 & 0 & 0 \\ 0 & 3 & 0 \\ 0 & 0 & -4 \end{bmatrix} \quad \text{gives the hyperbola} \quad -2x_1^2 + 3x_2^2 = 4,$$

$$\begin{bmatrix} 0 & 0 & -2 \\ 0 & 1 & 0 \\ -2 & 0 & 0 \end{bmatrix} \quad \text{gives the parabola} \quad x_2^2 = 4x_1.$$

Some polarities do not have self-conjugate points. For example, if

$$B = \begin{bmatrix} 1 & 0 & 0 \\ 0 & 1 & 0 \\ 0 & 0 & 1 \end{bmatrix},$$

then

$$b(x, y) = x_1 y_1 + x_2 y_2 + x_3 y_3 = \langle x, y \rangle,$$

and $b(x, y) = 0$ if and only if $x_1^2 + x_2^2 + x_3^2 = 0$. The fact the $\bar{b}$ has no self-conjugate points translates to the fact that no line can be perpendicular to itself in $\mathbf{E}^3$.

A polarity also induces a relation among the lines of $\mathbf{P}^2$. Two lines are said to be *conjugate* if the pole of one lies on the other. A line that passes through its own pole is said to be *self-conjugate*.

Theorem 21. *Let P and Q be points of $\mathbf{P}^2$ with respective polar lines p and q. Then P lies on q if and only if Q lies on p.*

Proof: Let $P = \pi x$ and $Q = \pi y$. Then P lies on q if and only if $b(x, y) = 0$. By symmetry this is also the condition for Q to lie on p. □

Theorem 22. *Let P and Q be self-conjugate points of $\mathbf{P}^2$. Then $\overleftrightarrow{PQ}$ cannot be a self-conjugate line.*

Proof: Let p and q be the respective polar lines of P and Q. The lines p and q are distinct because P and Q are distinct. Let R be the point where p and q intersect. This point is not on $\overleftrightarrow{PQ}$. Because R is conjugate to both P and Q, its polar line r must pass through both P and Q; that is, $r = \overleftrightarrow{PQ}$. The line r is not self-conjugate because it does not pass through R. $\quad\square$

Theorem 23. *A line contains exactly one self-conjugate point if and only if it is a self-conjugate line.*

Proof: Let ℓ be a line with exactly one self-conjugate point $P = \pi x$. Let $Q = \pi y$ be any other point of ℓ. Then for any real number λ,

$$b(x + \lambda y, x + \lambda y) = b(x, x) + 2\lambda b(x, y) + \lambda^2 b(y, y)$$
$$= \lambda(2b(x, y) + \lambda b(y, y)).$$

Because there is only one self-conjugate point on ℓ, we must have $b(x, y) = 0$. Otherwise, one could solve the equation for a nonzero value of λ. Because the equation $b(x, y) = 0$ holds for all x with πx on ℓ, the pole of ℓ is πy. Hence, ℓ is a self-conjugate line.

Conversely, if a line is self-conjugate, its pole is self-conjugate. By Theorem 22 the line can have no other self-conjugate points. $\quad\square$

Definition. *Let $\bar{b}$ be a polarity defining a conic $\mathscr{C}$. A line that is self-conjugate with respect to $\bar{b}$ is called a* tangent *to the conic $\mathscr{C}$. The pole of this line is called the* point of contact. *(See Figure 5.11 in which ℓ and m are tangents having respective points of contact L and M.)*

Corollary. *A line meets a conic in at most two points.*

Proof: This follows from considering a quadratic function of the type occurring in Theorem 23. $\quad\square$

Definition. *A line that meets a conic twice is called a* secant.

Cross products

Conjugacy with respect to a polarity is a generalization of the theory of perpendicularity with respect to an inner product. We recall that in order to find a vector in $\mathbf{E}^3$ that is perpendicular to two given vectors, we construct the cross product.

If u and v are vectors in $\mathbf{R}^3$, there is a unique vector w in $\mathbf{R}^3$ such that, for all $z \in \mathbf{R}^3$,

$$b(w, z) = \sqrt{|\det B|}\,\det(z, u, v).$$

Here we may compute the right side by writing z, u, and v as column vectors and taking the determinant of the resulting 3×3 matrix.

We call w the cross product (of u and v) with respect to b and write $w = u \times_b v$ or simply $w = u \times v$ if b is clear from the context.

Clearly, the formulas

$$b(u \times v, w) = b(u, v \times w)$$

and

$$b(u, u \times v) = b(v, u \times v) = 0$$

are true. Thus the cross product is a device for computing poles of lines. The following proposition is obvious.

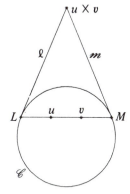

Theorem 24. *Let πu and πv be points in $\mathbf{P}^2$. Then the line joining πu and πv has pole $\pi(u \times v)$. (Again see Figure 5.11.)*

Figure 5.11 Conic; tangents, pole, and polar.

Definition *A triangle $\triangle PQR$ of $\mathbf{P}^2$ is said to be self-polar if each vertex is the pole of the side opposite it. Any self-polar triangle gives rise to a basis $\{e_1, e_2, e_3\}$ of $\mathbf{R}^3$ such that $b(e_i, e_j) = 0$ for $i \neq j$ and $b(e_i, e_i) = \pm 1$. Such a basis is said to be orthonormal with respect to b.*

Theorem 25.
i. *Let $\{e_1, e_2, e_3\}$ be orthonormal with respect to b. Then, after replacing e_3 by its negative if necessary, we have*

$$e_1 \times e_2 = b(e_3, e_3)e_3,$$
$$e_2 \times e_3 = b(e_1, e_1)e_1,$$
$$e_3 \times e_1 = {}^{\backprime}b(e_2, e_2)e_2.$$

ii. *For a given b the number of occurrences of -1 among the $b(e_i, e_i)$ is independent of the choice of orthonormal basis.*

Definition. *Let b be a nondegenerate, bilinear, symmetric function. Suppose that $\{e_i\}$ is a basis orthonormal with respect to b. Suppose that $+1$ occurs r times and -1 occurs s times among the $b(e_i, e_i)$. Then the ordered pair (r, s) is called the signature of b.*

The following (vector triple product) formula is indispensable for computation.

Theorem 26. $(u \times v) \times w = (-1)^s(b(u, w)v - b(v, w)u)$, *where the signature of b is* (r, s).

Proof: Choose a basis of the type used in Theorem 25. Then

$$(e_1 \times e_2) \times e_2 = b(e_3, e_3)e_3 \times e_2 = -b(e_3, e_3)b(e_1, e_1)e_1,$$

$$(-1)^s(b(e_1, e_2)e_2 - b(e_2, e_2)e_1) = (-1)^s(-1)b(e_2, e_2)e_1.$$

These are equal if and only if

$$b(e_1, e_1)b(e_3, e_3) = (-1)^s b(e_2, e_2);$$

that is,

$$b(e_1, e_1)b(e_2, e_2)b(e_3, e_3) = (-1)^s.$$

The other combinations can be checked similarly. □

EXERCISES

1. Prove Theorem 3.

2. Let $x = (1, 0, 0)$, $y = (1, 1, 0)$, $z = (1, 0, 1)$, $w = (1, 1, 1)$. Let ℓ be the line joining πx and πy, and let m be the line joining πz and πw. Find $\ell \cap m$.

3. Draw diagrams illustrating the various possibilities in Theorem 8.

4. Draw a diagram illustrating Theorem 9.

5. Draw a diagram illustrating Theorem 10.

6. Pappus' theorem yields many distinct results in $\mathbf{E}^2$ depending on the position of ℓ_∞. State as many of these results as you can.

7. Let ℓ and ℓ' be distinct lines, and let C be a point not on either line. The *perspectivity* $[C; \ell \to \ell']$ is the mapping α that sends each point $P \in \ell$ to the intersection of $\overleftrightarrow{PC}$ with ℓ'.
   i. Verify that the mapping α is well-defined.
   ii. Verify that α is a bijection with exactly one fixed point.
   iii. Verify that α^{-1} is a perspectivity.
   iv. Show that the composition of two perspectivities need not be a perspectivity.
   v. Given four distinct points P, Q, P', and Q', prove that there is a unique perspectivity taking P to P' and Q to Q'.

8. A *projectivity* is a composition of finitely many perspectivities. Each projectivity relates a pair of (not necessarily distinct) lines. For each line ℓ prove that the set of all projectivities that take ℓ to itself is a group. With respect to an appropriate choice of homogeneous coordinates, find a matrix representation for this group.

9. Let P, Q, and R be distinct points on a line ℓ, and let P', Q', and R' be distinct points on a line ℓ'. Prove that there is a unique projectivity sending P to P', Q to Q', and R to R'.

10. If ℓ and ℓ' are distinct in Exercise 9, show that the required projectivity may be expressed as the product of two perspectivities.

11. Show that any projectivity is the product of three or fewer perspectivities.

12. Let A, B, C, and D be four collinear points. Show that there is a unique projectivity that interchanges A and B and also interchanges C and D.

13. Show that a projectivity relating distinct lines is a perspectivity if and only if it has a fixed point.

14. Classify the projectivities of a given line ℓ in terms of their fixed point behavior.

15. Prove that a projective collineation that leaves fixed four points, no three of which are collinear, must be the identity. (*Hint*: Choose ℓ_∞ to be one of the fixed lines, and apply Theorem 2.2 to $\mathbf{P}^2 - \ell_\infty$.)

16. Prove the uniqueness part of Theorem 11 – there is only one projective collineation relating two specified quadrangles.

17. Prove that every projective collineation is of the form $\tilde{A}$ for some $A \in \mathbf{GL}(3)$.

18. The fixed lines of a projective collineation $\tilde{A}$ can be found by computing the eigenvectors of A^t. Justify this statement and use it to prove Theorem 12.

19. Prove Theorem 13.

20. If T is a projective collineation, prove that the restriction of T to one line ℓ is a projectivity. Prove also that every projectivity arises in this way.

21. Prove Theorem 14.

22. Show that the transformation of Theorem 14 preserves the relationships $(a - 1)x_1 + px_3 = 0$ and $(c - 1)x_2 + qx_3 = 0$, in addition to preserving the line $x_3 = 0$. Thus, unless $a = c = 1$, there is an additional fixed line. Use this to prove Theorem 15.

23. Prove Theorem 16 and its corollary.

24. Prove Theorem 17.

25. Prove Theorem 18.

26. A perspective collineation induces a projectivity on any fixed line. Discuss the fixed point behavior of such a projectivity.

27. Show that the set of perspective collineations with a given axis and

center (with the identity thrown in) is a group. Do these groups have any finite subgroups?

28. i.   Verify the remarks following Theorem 18.
     ii. Show that the harmonic homologies are those having $a = -1$ and $q = 0$ in Theorem 18.
     iii. Prove that the only projective collineations that are involutions are the harmonic homologies.

29. When ℓ_∞ is taken to be the axis of a harmonic homology, what affine transformation of $\mathbf{P}^2 - \ell_\infty$ results?

30. i.   Prove that there is a unique harmonic homology with a given center and axis.
     ii. If α is a harmonic homology with center P and axis ℓ and β is a harmonic homology with center Q and axis m, prove that $\alpha\beta = \beta\alpha$ if and only if Q lies on ℓ and P lies on m.

31. Prove Theorem 19.

32. Prove Theorem 20.

33. Let $\tilde{b}$ be a polarity, and let ℓ be a non-self-conjugate line. For each $X \in \ell$, let $\alpha(X)$ be the point where the polar line of X intersects ℓ. Prove that α is a projectivity. Show further that $\alpha^2 = I$; that is, α is an involution.

34. Prove Theorem 23.

35. Prove Theorem 25.

Distance geometry on P² 6

Distance and the triangle inequality

So far we have discussed only the incidence structure of the projective plane. We now introduce a distance function.

Definition. *For P and Q in* **P**² *define*

$$d(P, Q) = \cos^{-1}|\langle x, y\rangle|,$$

where x and y are points of **S**² *and* $\pi x = P$, $\pi y = Q$.

Remark: Because of the absolute value sign, the distance is well-defined. The distance between $\{x, -x\}$ and $\{y, -y\}$ is the spherical distance between the closest representatives. Also, all distances are $\leqslant \pi/2$.

Theorem 1. *If P, Q, and R are points of* **P**², *then*

$$d(P, Q) + d(Q, R) \geqslant d(P, R).$$

Proof: Let r, p, and q be the respective distances. Choose representatives P', Q', and R' so that $\langle P', R'\rangle \geqslant 0$ and $\langle Q', R'\rangle \geqslant 0$. As in Theorem 4.8,

$$|P' \times Q'| = \sin r, \quad |R' \times Q'| = \sin p,$$

$$|P' \times Q'||R' \times Q'| \geqslant \langle P' \times Q', Q' \times R'\rangle$$
$$= -\langle P', R'\rangle + \langle Q', R'\rangle\langle Q', P'\rangle.$$

If $\langle P', Q'\rangle \geqslant 0$, then we get the inequalities

$$\sin r \sin p \geqslant \cos p \cos r - \cos q,$$

$$\cos q \geqslant \cos p \cos r - \sin r \sin p,$$

$$\cos q \geqslant \cos (p + r),$$

$$q \leqslant p + r.$$

Equality in this case would imply that

$$[Q' \times R'] = [P' \times Q'].$$

This means that P', Q', and R' are collinear on $\mathbf{S}^2$ and, hence, that P, Q, and R are collinear on $\mathbf{P}^2$.

If $\langle P', Q' \rangle \leq 0$, then use the Cauchy–Schwarz inequality in the form

$$|P' \times Q'||R' \times Q'| \geq \langle P', R' \rangle - \langle Q', R' \rangle \langle Q', P' \rangle, \qquad (6.1)$$

which yields

$$\sin r \sin p \geq \cos q + \cos p \cos r,$$

$$\cos(\pi - q) \geq \cos(p + r),$$

$$\pi - q \leq p + r.$$

But $q \leq \pi/2$, so that $q \leq \pi - q \leq p + r$. Equality can occur in this case only if $q = \pi/2 \leq p + r$. As before, P', Q', and R' will be collinear. $\qquad \square$

Remark: In this proof we have shown not only that the triangle inequality holds on $\mathbf{P}^2$ but also the familiar notion that equality cannot occur unless the three points in question are collinear. However, something unfamiliar also pops up here. Not every triple of collinear points satisfies the equality. The situation is as follows.

Theorem 2. *Three points of $\mathbf{P}^2$ are collinear if and only if they can be named P, Q, and R in such a way that either*

or
   i. $d(P, Q) + d(Q, R) = d(P, R)$
   ii. $d(P, Q) + d(Q, R) + d(P, R) = \pi.$

Proof: Suppose that we are given three collinear points for which (i) does not hold. Let e_3 be a pole of the line of $\mathbf{S}^2$ determined by the three given points, and let e_1 be a representative of one of the points, say P. Then we may choose θ and ϕ with $0 \leq \theta, \phi \leq \pi/2$ such that the other two points are

$$Q = \pi((\cos \phi)e_1 + (\sin \phi)e_2)$$

and

$$R = \pi((\cos \theta)e_1 \pm (\sin \theta)e_2),$$

where $\{e_1, e_2, e_3\}$ is an orthonormal basis.

   Now

$$d(P, Q) = \cos^{-1}(\cos \phi) = \phi,$$

$$d(P, R) = \cos^{-1}(\cos \theta) = \theta,$$

and

$$d(Q, R) = \cos^{-1}|\cos(\phi \pm \theta)|.$$

The minus sign cannot occur because it would imply that $d(Q, R) = |\phi - \theta|$, and, hence, an equation of the form (i) would be satisfied. Similarly, if $\phi + \theta \leq \pi/2$, we would have $d(Q, R) = \phi + \theta$, another version of (i). Thus, we must conclude that $\pi/2 < \phi + \theta < \pi$, so that

$$
\begin{aligned}
d(Q, R) &= \cos^{-1}(-\cos(\phi + \theta)) = \cos^{-1}(\cos(\pi - (\theta + \phi))) \\
&= \pi - (\theta + \phi) = \pi - d(P, Q) - d(P, R).
\end{aligned}
$$

Conversely, if (i) holds, we showed in the proof of Theorem 1 that the points must be collinear. If (ii) holds, we have $p + r = \pi - q$, so that $\cos(\pi - q) = \cos(p + r)$. By the same algebra as in Theorem 1, (6.1) becomes an equality, and the three points are collinear. $\qquad\square$

Isometries

Definition. *A map* $T: \mathbf{P}^2 \to \mathbf{P}^2$ *is called an* isometry *if* $d(P, Q) = d(TP, TQ)$ *for all P and Q in* $\mathbf{P}^2$.

Theorem 3. *Let T be an isometry. If P, Q, and R are collinear, then TP, TQ, and TR are collinear.*

Proof: Let P, Q, and R be collinear points. Let P' be the unique point on this line such that $d(P, P') = \pi/2$. Then

$$
d(P, Q) + d(Q, P') = d(P, P') = \frac{\pi}{2}.
$$

Hence,

$$
d(TP, TQ) + d(TQ, TP') = d(TP, TP') = \frac{\pi}{2}.
$$

By the previous theorem, TQ must lie on the line determined by TP and TP'. Similarly, TR must lie on this line. $\qquad\square$

The isometries of $\mathbf{P}^2$ are closely related to the isometries of $\mathbf{S}^2$.

Theorem 4. *Let $T: \mathbf{P}^2 \to \mathbf{P}^2$ be an isometry. Then there exists a unique $A \in \mathbf{SO}(3)$ such that $T = \bar{A}$.*

Proof: Choose e_1, e_2, and e_3 on $\mathbf{S}^2$ such that

$$
T\pi\varepsilon_i = \pi e_i \quad \text{for each } i.
$$

Then

$$
d(T\pi\varepsilon_i, T\pi\varepsilon_j) = d(\pi\varepsilon_i, \pi\varepsilon_j) = \cos^{-1}|\langle \varepsilon_i, \varepsilon_j \rangle|.
$$

But $d(\pi e_i, \pi e_j) = \cos^{-1}|\langle e_i, e_j \rangle|$, and so $|\langle \varepsilon_i, \varepsilon_j \rangle| = |\langle e_i, e_j \rangle|$. If $i \neq j$, then $\langle e_i, e_j \rangle = 0$. Otherwise, $\langle e_i, e_j \rangle = 1$. Thus, $\{e_i\}$ is an orthonormal basis of $\mathbf{R}^3$. Let A be the orthogonal matrix such that $A\varepsilon_i = e_i$ for each i. Then $\tilde{A}$ is an isometry of $\mathbf{P}^2$, and $\tilde{A}^{-1}T$ leaves $\pi\varepsilon_1$, $\pi\varepsilon_2$, and $\pi\varepsilon_3$ fixed. Let $M = \pi(\varepsilon_1 + \varepsilon_2 + \varepsilon_3)$ and write

$$\tilde{A}^{-1}TM = \pi(k_1\varepsilon_1 + k_2\varepsilon_2 + k_3\varepsilon_3),$$

where k_1, k_2, and k_3 are some numbers, with $k_1^2 + k_2^2 + k_3^2 = 1$. We claim that the $|k_i|$ are all equal. To see this, note that

$$d(\tilde{A}^{-1}TM, \tilde{A}^{-1}T\pi\varepsilon_i) = \cos^{-1}|k_i|.$$

But $d(M, \pi\varepsilon_i) = \cos^{-1}(1/\sqrt{3})$ and, hence, $|k_i| = 1/\sqrt{3}$ for all i. Let

$$B = \sqrt{3}\begin{bmatrix} k_1 & 0 & 0 \\ 0 & k_2 & 0 \\ 0 & 0 & k_3 \end{bmatrix}.$$

Then

$$\tilde{B}\tilde{A}^{-1}T\pi\varepsilon_i = \pi\varepsilon_i \quad \text{for each } i,$$

and since each k_i^2 is $1/3$, $\tilde{B}\tilde{A}^{-1}TM = M$. $\qquad\square$

We will next prove that any isometry that leaves each ε_i and $\pi(\varepsilon_1 + \varepsilon_2 + \varepsilon_3)$ fixed must be the identity. Assuming this for the moment, we get

$$\tilde{B}\tilde{A}^{-1}T = I;$$

that is,

$$T = (\tilde{A}^{-1})^{-1}\tilde{B}^{-1} = \tilde{A}\tilde{B}^{-1} = \widetilde{AB}^{-1}.$$

But $B^{-1} = B$. Hence, $T = \widetilde{AB}$. Because AB is orthogonal, there is a unique member of $\mathbf{SO(3)}$ that determines the same isometry.

Theorem 5. *If T is an isometry of $\mathbf{P}^2$ that leaves fixed each $\pi\varepsilon_i$ and $\pi(\varepsilon_1 + \varepsilon_2 + \varepsilon_3)$, then T is the identity.*

Proof: Let us work in the homogeneous coordinate system determined by ε_1, ε_2, and ε_3. Then $(1, 0, 0)$, $(0, 1, 0)$, $(0, 0, 1)$, and $(1, 1, 1)$ are fixed points. We first check that all points on the line joining $(1, 0, 0)$ and $(0, 1, 0)$ are fixed. A typical such point is $x = (\cos \alpha, \sin \alpha, 0)$, where $0 < \alpha < \pi$. Write $Tx = (\cos \beta, \sin \beta, 0)$, where $0 < \beta < \pi$. Then

$$d(T\pi\varepsilon_1, Tx) = d(\pi\varepsilon_1, x),$$

$$\cos^{-1}|\cos \beta| = \cos^{-1}|\cos \alpha|;$$

that is, $|\cos \beta| = |\cos \alpha|$. Thus, $\beta = \alpha$ or $\beta = \pi - \alpha$.
 If $\beta = \alpha$, we are finished. If $\beta = \pi - \alpha$, then

$$Tx = \pi(-\cos \alpha, \sin \alpha, 0).$$

Let $M = (1, 1, 1)$. Then

$$d(x, M) = \cos^{-1} \left| \frac{1}{\sqrt{3}} (\cos \alpha + \sin \alpha) \right|,$$

$$d(Tx, M) = \cos^{-1} \left| \frac{1}{\sqrt{3}} (-\cos \alpha + \sin \alpha) \right|.$$

This is impossible unless $\alpha = \pi/2$, in which case $\alpha = \beta$ anyway. Similarly, we can show that all points on the sides of the triangle of reference Δ are fixed. Now each line of $\mathbf{P}^2$ contains at least two fixed points because it intersects Δ at least twice. Therefore, every line is fixed, and, hence, every point is a fixed point. $\qquad\square$

Motions

Let ℓ be a line of $\mathbf{S}^2$. Then the *reflection* in the line $\pi\ell$ is the isometry of $\mathbf{P}^2$ defined by

$$\Omega_{\pi\ell} = \tilde{\Omega}_\ell.$$

Theorem 6. $\Omega_{\pi\ell}$ *leaves fixed every point on $\pi\ell$ and the pole of $\pi\ell$. No other points are fixed.*

Proof: Choose an orthonormal basis of $\mathbf{E}^3$ with respect to which

$$\Omega_\ell = \begin{bmatrix} -1 & 0 & 0 \\ 0 & 1 & 0 \\ 0 & 0 & 1 \end{bmatrix}.$$

Now $\Omega_\ell x = x$ if and only if x lies on the line joining $(0, 1, 0)$ and $(0, 0, 1)$. Also $\Omega_\ell x = -x$ if and only if $x = (1, 0, 0)$ or $(-1, 0, 0)$. Thus, the fixed points of Ω_ℓ are as claimed. $\qquad\square$

 Let ℓ be a line of $\mathbf{P}^2$, and let ξ be its pole. Then the product of two reflections $\Omega_m\Omega_n$, where m and n pass through ξ, is called a *rotation about* ξ. Because line goes through ξ if and only if it is perpendicular to ℓ, we also call $\Omega_m\Omega_n$ a *translation along* ℓ, and we call $\Omega_m\Omega_n\Omega_\ell$ a *glide reflection*. If $m \perp n$, then $\Omega_m\Omega_n$ is called a *half-turn*.

Theorem 7. *The fixed lines of a reflection are the line of fixed points and all lines perpendicular to this line.*

Proof: A line is fixed if and only if its pole is a fixed point. □

Theorem 8. *A rotation other than a half-turn or the identity has a unique fixed point and a unique fixed line. The point is the pole of the line.*

Proof: Let $\bar{A}$ be a typical rotation of $\mathbf{P}^2$, where $A \in \mathbf{SO(3)}$. Then we will have solved the problem if we can find all those points $x \in \mathbf{S}^2$ such that $Ax = \pm x$. But this calculation was done in finding the fixed lines of a rotation of $\mathbf{S}^2$. If x is a fixed point of A, then πx is the unique fixed point of $\bar{A}$, and the line whose pole is πx is the unique fixed line of $\bar{A}$. □

Theorem 9.
i. *Every reflection is a half-turn, and every half-turn is a reflection.*
ii. *Every glide reflection is a rotation.*

Corollary. *Every isometry of $\mathbf{P}^2$ is a rotation.*

Remark: It is easy to show that the three reflections theorem and the representation theorems for rotations and translations (Theorems 4.15, 16, 19, and 20) hold in $\mathbf{P}^2$ (Exercise 6).

Elliptic geometry

The geometry of $\mathbf{P}^2$ is traditionally called elliptic geometry. So far, we have discussed its incidence properties, defined the notion of distance, and classified the isometries. We have seen that elliptic geometry is a simplification of spherical geometry.

Definition. *A* segment *in $\mathbf{P}^2$ is a set of the form $\pi\jmath$, where $\jmath$ is a minor segment in $\mathbf{S}^2$. The length of $\pi\jmath$ is the length of $\jmath$. The end points of $\pi\jmath$ are the images by π of the end points of $\jmath$.*

Theorem 10.
i. *Each pair $\{A, B\}$ of points in $\mathbf{P}^2$ is the end point set of two segments. The union of these segments is the line $\overleftrightarrow{AB}$, and their intersection is $\{A, B\}$.*
ii. *For a segment of length L with end points A and B, we have $d(A, B) = L$ if $L \leq \pi/2$. Otherwise, $d(A, B) = \pi - L$.*

Definition. *A* ray *is a segment of length $\pi/2$ with one end point removed. The remaining endpoint is called the* origin *of the ray.*

Remark: The definitions of ray in Euclidean, spherical, and elliptic geometry may seem at first to have little in common. There is a unifying idea, however. Starting at the origin of the ray, we move in a particular direction as long as the path we have traced out is the shortest path to this origin. If this continues forever, as in the Euclidean case, the ray continues forever. On the sphere, however, once we reach the point antipodal to the ray's origin, we lose this uniqueness. In differential geometry the point where this happens is called a *cut point*. In $\mathbf{P}^2$ we reach a cut point at distance $\pi/2$.

Theorem 11. *Let P and Q be points with $d(P, Q) < \pi/2$. Then there is a unique ray with origin P that contains Q. We denote this ray by $\overrightarrow{PQ}$.*

The definition of angle in elliptic geometry is the same as our previous definitions. The radian measure of an angle $\measuredangle PQR$ is determined by choosing a representative for Q, choosing the representatives for P and R closest to Q, and computing the radian measure of the spherical angle so determined.

The notion of half-plane does not occur in $\mathbf{P}^2$. One can, however, define the interior of an angle.

A *triangle* in elliptic geometry is a figure of the form $\pi\Delta$, where Δ is a spherical triangle.

Theorem 12. *If P, Q, and R are three noncollinear points of $\mathbf{P}^2$, there is a triangle having P, Q, and R vertices. The triangle is the union of three segments.*

Remark: Our treatment of elliptic geometry has been brief. Most of the questions we have studied in Euclidean and spherical geometry have analogues that can be studied in the elliptic setting. Some of these are explored in the exercises.

EXERCISES

1. Find the distance $d(P, Q)$, where $P = (-1, 0, 1)$ and $Q = (1, 1, 0)$ in homogeneous coordinates with respect to $\{\epsilon_1, \epsilon_2, \epsilon_3\}$.

2. i. Prove that every pair of distinct lines in $\mathbf{P}^2$ has a common perpendicular. Only a slight modification of your proof of Theorem 4.10 is required.
    ii. Find the common perpendicular to the lines $x_1 + 2x_2 = 0$ and $2x_2 - x_3 = 0$.

3. i. Prove the projective version of Theorem 4.11 concerning erecting and dropping perpendiculars.

   ii. Is the foot of the perpendicular the point on ℓ closest to P?

4. Let $A: \mathbf{R}^3 \to \mathbf{R}^3$ be linear. Then define $\tilde{A}\pi x = \pi Ax$ so that $\tilde{A}$ maps $\mathbf{P}^2 \to \mathbf{P}^2$. Under what conditions on A will $\tilde{A}$ be an isometry? Illustrate using the matrix

$$\begin{bmatrix} 1 & 1 & 0 \\ 0 & 0 & 1 \\ 1 & -1 & 0 \end{bmatrix}.$$

5. Prove Theorem 9 and its corollary.

6. Verify the remark following Theorem 9 (that the three reflections and representation theorems hold in $\mathbf{P}^2$).

7. i. Given three nonconcurrent lines α, β, and γ, show how to find a point P and a line ℓ such that $\Omega_\alpha\Omega_\beta\Omega_\gamma = \Omega_\ell H_P$.

   ii. If $\Omega_\ell H_P = \Omega_m H_Q$, what relationships must hold among ℓ, m, P, and Q?

8. Let P and Q be distinct points. Find $\mathscr{S}(\{P, Q\})$.

9. Under what conditions will two rotations about distinct points commute?

10. Let P, Q, and R be mutually perpendicular points of $\mathbf{S}^2$. Show that there are four isometries T of $\mathbf{P}^2$ that leave πP, πQ, and πR fixed. (*Hint*: Choose an appropriate orthonormal basis and compute the possible forms of the matrix of T.)

11. Find the symmetry group of the figure in $\mathbf{P}^2$ formed by two perpendicular lines.

12. Suppose that an isometry T of $\mathbf{P}^2$ has three concurrent fixed lines. Show that T must be a half-turn.

13. Prove Donkin's theorem: Let PQR be a triangle. Let α, β, and γ be rotations (translations) that take P to Q, Q to R, and R to P, respectively. Then $\gamma\beta\alpha$ is the identity.

14. Prove Theorem 10.

15. Prove Theorem 11.

16. Let P and Q be points with $d(P, Q) < \pi/2$. Prove that $\overrightarrow{PQ} \cap \overrightarrow{QP}$ is the segment with end points P and Q and length $d(P, Q)$.

17. Prove that the notion of radian measure for angles in $\mathbf{P}^2$ is well-defined.

18. What happens to Theorem 4.41 in $\mathbf{P}^2$?

19. i. Propose a definition for the perpendicular bisector of a segment in $\mathbf{P}^2$.

ii. Define the midpoint of a segment in such a way that each segment has a unique midpoint.

20. Prove that there are exactly two reflections that interchange a given pair of lines in $\mathbf{P}^2$.

21. Let ℓ be a line of $\mathbf{P}^2$, and let P and Q be any points not on ℓ. Show that there is a segment joining P and Q that does not meet ℓ. (This is why we do not attempt to define the notion of half-plane in $\mathbf{P}^2$.)

22. Define the interior of an angle in $\mathbf{P}^2$. Does the crossbar theorem hold?

23. Define the interior of a triangle in $\mathbf{P}^2$. Show that $\mathbf{P}^2$ may be regarded as the union of four equilateral triangles (and their interiors). Each triangle should have three right angles.

24. Prove Theorem 12. Is the triangle unique?

25. Prove that if X is a point in the interior of a triangle Δ, there is a segment containing X whose end points are on Δ.

26. Prove that the perpendicular bisectors of the three sides of a triangle are concurrent. In light of Exercise 18, what further results can be obtained?

27. Prove that the congruence theorems for spherical triangles are valid in $\mathbf{P}^2$ as well (Theorems 55–57 of Chapter 4).

28. i. Show that the finite groups of isometries of $\mathbf{P}^2$ may be identified with those listed in Theorem 4.58.
    ii. For each such group find a figure in $\mathbf{P}^2$ of which it is the symmetry group.

7

The hyperbolic plane

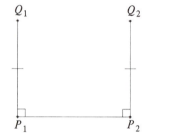

Figure 7.1 In hyperbolic geometry, $d(Q_1, Q_2) > d(P_1, P_2)$.

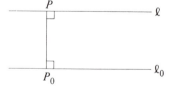

Figure 7.2 In hyperbolic geometry, ℓ_0 and ℓ do not meet.

Introduction

The projective plane provides one alternative to Euclidean geometry. A second alternative is explored in this chapter.

The three geometries are contrasted in the following example: Take a segment P_1P_2 as shown in Figure 7.1. Erect equal segments P_1Q_1 and P_2Q_2 perpendicular to P_1P_2.

In $\mathbf{E}^2$ the segment Q_1Q_2 will have length equal to that of P_1P_2. However, in $\mathbf{P}^2$, the length of Q_1Q_2 will be less than that of P_1P_2. In $\mathbf{H}^2$ we shall see that Q_1Q_2 will be longer than P_1P_2.

This construction is also related to the question of parallelism. Let ℓ_0 be a line, and let P be a point not on ℓ_0. Drop a perpendicular PP_0 from P to ℓ_0, and let ℓ be the line through P perpendicular to PP_0. (See Figure 7.2.)

In $\mathbf{E}^2$, ℓ will be parallel to ℓ_0. In $\mathbf{P}^2$, ℓ will meet ℓ_0. In $\mathbf{H}^2$ it will turn out that ℓ does not meet ℓ_0.

We will now proceed to construct the geometry $\mathbf{H}^2$. It will again consist of "points" and "lines" with a "distance" function defined for each pair of points. As in the case of $\mathbf{E}^2$ and $\mathbf{P}^2$, we find that isometries of $\mathbf{H}^2$ are generated by reflections and satisfy the three reflections theorems.

Algebraic preliminaries

Our model of spherical geometry was a certain subset of $\mathbf{R}^3$, and the usual inner product of $\mathbf{R}^3$ played an important role. Our model of hyperbolic geometry will also be a subset of $\mathbf{R}^3$. However, the bilinear form on which hyperbolic geometry is based is defined by

$$b(x, y) = x_1y_1 + x_2y_2 - x_3y_3$$

(see also Chapter 6). A function of this type is used in Einstein's special theory of relativity. (See Frankel [15] or Taylor–Wheeler [29].) This explains some of the terms used in discussing its properties.

Definition. *A nonzero vector $v \in \mathbf{R}^3$ is said to be*

i. *spacelike if $b(v, v) > 0$. If $b(v, v) = 1$, it is a unit spacelike vector. An example is ε_1.*
ii. *timelike if $b(v, v) < 0$. If $b(v, v) = -1$, it is a unit timelike vector. An example is ε_3.*
iii. *lightlike if $b(v, v) = 0$. An example is $\varepsilon_1 - \varepsilon_3$.*

We use the notation $|v|$ for the "length" of a vector v (i.e., $|v| = |b(v, v)|^{1/2}$). Unit vectors satisfy $|v| = 1$.

In this chapter we use the term "orthonormal" to mean orthonormal with respect to b. Note that $\{\varepsilon_1, \varepsilon_2, \varepsilon_3\}$ is orthonormal.

Theorem 1.
i. *Every orthonormal set of three vectors is a basis for $\mathbf{R}^3$.*
ii. *Every orthonormal basis has two spacelike vectors and one timelike vector.*
iii. *For every orthonormal pair $\{u, v\}$ of vectors, $\{u, v, u \times v\}$ is an orthonormal basis. (The cross product is taken with respect to b.)*
iv. *For every unit spacelike or unit timelike vector v, there is an orthonormal basis containing v.*

Proof:
i. We need only show that an orthonormal set is linearly independent. If an equation of the form

$$0 = \lambda_1 e_1 + \lambda_2 e_2 + \lambda_3 e_3$$

holds, where $\{e_1, e_2, e_3\}$ is orthonormal, then for each i,

$$0 = b(0, e_i) = \lambda_i b(e_i, e_i)$$

implies that $\lambda_i = 0$.
ii. First note that all three vectors cannot be spacelike. In fact, if all $b(e_i, e_i)$ are equal and

$$x = \sum_{i=1}^{3} x_i e_i,$$

we have

$$b(x, x) = \sum_{i=1}^{3} x_i^2 b(e_i, e_i).$$

This would imply that all vectors are spacelike. Similarly, if all the e_i were timelike, every vector in $\mathbf{R}^3$ would be timelike. We conclude that any orthonormal basis has at least one spacelike vector and one timelike vector.

Let $\{e_1, e_2, e_3\}$ be an orthonormal basis. Suppose that e_1 is spacelike

and e_3 is timelike. Then $(e_1 \times e_3) \times e_2 = 0$, so that e_2 is a multiple of $e_1 \times e_3$. Further,

$$b(e_1 \times e_3, e_1 \times e_3) = -b(e_1, e_1)\, b(e_3, e_3) = 1,$$

so that $e_1 \times e_3$ (and hence e_2) is spacelike.

iii. We note that

$$b(u \times v, u \times v) = -b(u, u)\, b(v, v) = \pm 1,$$

and, hence, $\{u, v, u \times v\}$ is orthonormal.

iv. Suppose that v is spacelike. Let w be any unit timelike vector (e.g., $\varepsilon_3 = (0, 0, 1)$). If $b(v, w) = 0$, we can use $\{v, w, v \times w\}$ as our basis. If not, choose $\bar{u} = v + \lambda w$, where $\lambda = -1/b(v, w)$. Then

$$\begin{aligned} b(\bar{u}, \bar{u}) &= 1 + 2\lambda b(v, w) - \lambda^2 \\ &= 1 - 2 - \lambda^2 = -(1 + \lambda^2). \end{aligned}$$

If we set

$$u = \frac{v + \lambda w}{\sqrt{1 + \lambda^2}},$$

then $\{u, v, u \times v\}$ is an orthonormal basis.

Suppose now that v is timelike. A similar construction, using a unit spacelike vector w, leads to an orthonormal basis $\{u, v, u \times v\}$, where $u = (v + \lambda w)/\sqrt{1 + \lambda^2}$ and $\lambda = 1/b(v, w)$. $\square$

Theorem 2.

i. *For any $x \in \mathbf{R}^3$,*

$$x = \sum_{i=1}^{3} b(x, e_i) b(e_i, e_i) e_i \tag{7.1}$$

if $\{e_1, e_2, e_3\}$ is an orthonormal basis.

ii. *Let v be a timelike vector. Suppose that $w \times v \neq 0$ and $b(v, w) = 0$. Then w is spacelike.*

The Cauchy–Schwarz inequality played an important role in $\mathbf{E}^2$ and $\mathbf{S}^2$. Here is the hyperbolic version.

Theorem 3. *Let ξ and η be spacelike vectors in $\mathbf{R}^3$ such that $\xi \times \eta$ is timelike. Then*

$$b(\xi, \eta)^2 < b(\xi, \xi) b(\eta, \eta). \tag{7.2}$$

Proof: Let P be a unit timelike vector in the direction $[\xi \times \eta]$. As in the proof of Theorem 1.4, we consider the function

$$f(t) = b(\xi + t\eta, \xi + t\eta).$$

Because $b(\xi + t\eta, P) = 0$ for all real values of t and $P \times (\xi + t\eta) \neq 0$, Theorem 2 applies, and $\xi + t\eta$ is spacelike. In other words, $f(t) > 0$ for all t and

$$b(\xi, \eta)^2 < b(\xi, \xi)b(\eta, \eta). \qquad \square$$

Remark: If we weaken the hypothesis to $b(\xi \times \eta, \xi \times \eta) \leqslant 0$, the conclusion becomes

$$b(\xi, \eta)^2 \leqslant b(\xi, \xi)b(\eta, \eta).$$

However, equality can occur even if ξ and η are not proportional. (See Exercise 2.)

There is a similar result for timelike vectors.

Theorem 4. *Let v and w be timelike vectors. Then*

$$b(v, w)^2 \geqslant b(v, v)b(w, w). \qquad (7.3)$$

Proof: By Theorem 2, $v \times w$ is spacelike or zero. Thus

$$b(v \times w, v \times w) \geqslant 0.$$

In other words,

$$b(v, v)b(w, w) - b(v, w)^2 \leqslant 0$$

with equality holding if and only if v and w are proportional. $\qquad \square$

Corollary. *If v and w are unit timelike vectors, then $|b(v, w)| \geqslant 1$. The "inner product" $b(v, w)$ is positive if and only if $b(v, \varepsilon_3)$ and $b(w, \varepsilon_3)$ have opposite signs.*

Proof: The first statement is immediate from the theorem. To prove the second, we introduce the following notation. Let $v = (p_1, p_2, r)$ and $w = (q_1, q_2, s)$. Consider $p = (p_1, p_2)$ and $q = (q_1, q_2)$ as vectors in $\mathbf{R}^2$. Then

$$b(v, w) = \langle p, q \rangle - rs.$$

Because $(|p| + |q|)^2 \geqslant 0$ with equality if and only if $p = q = 0$, we have

$$|p|^2 + |q|^2 \geqslant -2|p||q|.$$

Adding $1 + |p|^2|q|^2$ to each side yields

$$(1 + |p|^2)(1 + |q|^2) \geqslant (|p||q| - 1)^2.$$

But $|p|^2 - r^2 = -1$ and $|q|^2 - s^2 = -1$, so that

$$(|p\|q| - 1)^2 \leqslant r^2s^2. \tag{7.4}$$

Suppose now that r and s are both positive but $b(v, w)$ is also positive. Then $\langle p, q \rangle \geqslant 1 + rs$; that is, $\langle p, q \rangle - 1 \geqslant rs$. By the Cauchy–Schwarz inequality for $\mathbf{R}^2$, we get

$$|p\|q| - 1 \geqslant rs,$$

which is incompatible with (7.4). We conclude that $b(v, w)$ must be negative when r and s are positive. The conclusion now follows from the linearity of the function b. $\qquad\square$

Incidence geometry of $\mathbf{H}^2$

The hyperbolic plane $\mathbf{H}^2$ is defined as follows:

$$\mathbf{H}^2 = \{x \in \mathbf{R}^3 | x_3 > 0 \quad \text{and} \quad b(x, x) = -1\}.$$

Thus, as a set, $\mathbf{H}^2$ is just the upper half of a hyperboloid of two sheets.

Definition. *Let ξ be a unit spacelike vector. Then*

$$\ell = \{x \in \mathbf{H}^2 | b(\xi, x) = 0\}$$

is called the line *with unit normal (or pole) ξ.*

Remark: Like the situation in spherical geometry, a line of $\mathbf{H}^2$ is the intersection with $\mathbf{H}^2$ of a plane through the origin of $\mathbf{R}^3$. Not all planes through the origin meet $\mathbf{H}^2$. However, if ξ is timelike, it can be completed to a basis orthonormal with respect to b (Theorem 1). In particular, there are points $x \in \mathbf{H}^2$ such that $b(\xi, x) = 0$. We will now proceed to a detailed study of lines in hyperbolic geometry.

Theorem 5. *Let P and Q be distinct points of $\mathbf{H}^2$. Then there is a unique line containing P and Q, which we denote by $\overleftrightarrow{PQ}$.*

Proof: Apply Theorem 2(ii) with $v = P$ and $w = P \times Q$. The triple product formula shows that $P \times (P \times Q) \neq 0$ and, hence, that $P \times Q$ is spacelike. Let ξ be a unit vector in the direction $[P \times Q]$. Then the line whose unit normal is ξ must pass through P and Q. This is the only line through P and Q because the unit normal to any such line must be orthogonal to P and Q (with respect to b) and, hence, must be a multiple of $P \times Q$. $\qquad\square$

Just as in spherical geometry, the cross product is used to find the point of intersection of a pair of lines. However, if ξ and η are spacelike unit vectors, $\xi \times \eta$ need not be timelike, and therefore the lines may not intersect in $\mathbf{H}^2$. In fact, all three possibilities for $\xi \times \eta$ can occur. This is what makes $\mathbf{H}^2$ a richer incidence geometry than any we have studied previously.

Definition. *Let ℓ and m be two lines with respective unit normals ξ and η. We say that ℓ and m are*
  i. intersecting *lines if $\xi \times \eta$ is timelike,*
 ii. parallel *lines if $\xi \times \eta$ is lightlike,*
iii. ultraparallel *lines if $\xi \times \eta$ is spacelike.*

Theorem 6. *Intersecting lines have exactly one point in common. This point is the unique point of $\mathbf{H}^2$ that is a multiple of $\xi \times \eta$.*

Proof: Clearly, the point in question lies on both lines. If P is any other point that lies on both lines, then

$$P \times (\xi \times \eta) = -b(P, \eta)\xi + b(P, \xi)\eta = 0,$$

so that P is a multiple of $\xi \times \eta$ as required. $\qquad\square$

Remark: Neither parallel nor ultraparallel lines intersect.

Perpendicular lines

Definition. *Two lines with unit normals ξ and η are said to be* perpendicular *if $b(\xi, \eta) = 0$.*

Theorem 7. *If two lines are ultraparallel, there is a unique line γ that is perpendicular to both of them. Conversely, if two lines have a common perpendicular, they must be ultraparallel.*

Proof: Let ξ and η be unit normals of two ultraparallel lines. Let ζ be the unit (spacelike) vector that is a multiple of $\xi \times \eta$. Then $b(\xi, \zeta) = b(\eta, \zeta) = 0$, so the line with unit normal ζ is a common perpendicular to the two lines.

Conversely, if the two lines have a common perpendicular, its unit normal ζ is a spacelike vector satisfying $\zeta \times (\xi \times \eta) = 0$ and, thus, is a multiple of $\xi \times \eta$. This means that $\xi \times \eta$ is spacelike, and the lines are ultraparallel. $\qquad\square$

Theorem 8.
  i. *If ℓ and m are perpendicular lines of $\mathbf{H}^2$, then ℓ intersects m.*
 ii. *Let X be a point of $\mathbf{H}^2$ and ℓ a line of $\mathbf{H}^2$. Then there is a unique line through X perpendicular to ℓ.*

Proof:

i. Let ξ and η be unit normals to ℓ and m, respectively. Then $\{\xi, \eta, \xi \times \eta\}$ is an orthonormal basis by Theorem 1. Hence, $\xi \times \eta$ is timelike.

ii. Let ξ be a unit normal to ℓ. Let η be a unit vector proportional to $\xi \times X$. This is possible because $\xi \times X$, being a nonzero vector orthogonal to X, must be spacelike.

   The line m whose unit normal is η clearly passes through X but is perpendicular to ℓ. There is only one line with this property, because a unit normal to such a line must be orthogonal to ξ and X and, therefore, a multiple of their cross product. $\square$

Definition. *The point F where m intersects ℓ is called the* foot *of the perpendicular from X to ℓ (provided X is not on ℓ).*

Remark: In the next section we define distance between two points of $\mathbf{H}^2$. As in $\mathbf{E}^2$ we can use this to define

$$d(X, \ell) = d(X, F),$$

where F is the foot of the perpendicular from X to ℓ.

Pencils

Definition. *Let ℓ and m be a pair of distinct lines with respective unit normals ξ and η. Then the set $\mathcal{P}$ of lines whose unit normals ζ are orthogonal to $\xi \times \eta$ is called a* pencil *of lines. $\mathcal{P}$ is called a pencil of intersecting lines, a pencil of parallels, or a pencil of ultraparallels according to whether $\xi \times \eta$ is timelike, lightlike, or spacelike.*

Remark: At the moment this definition may look somewhat strange. Clearly, if $\xi \times \eta$ is timelike, then lines with unit normal ζ will be the lines passing through the point of intersection, as expected. If $\xi \times \eta$ is spacelike, the pencil will consist of all lines perpendicular to a certain line. However, it is not yet evident what the pencil looks like when $\zeta \times \eta$ is lightlike. When we look at $\mathbf{H}^2$ as a subset of $\mathbf{P}^2$, we will get a more concrete interpretation for $\xi \times \eta$ and the associated pencils.

Remark:

i. The set of all lines of $\mathbf{H}^2$ perpendicular to a certain line of $\mathbf{H}^2$ is a pencil of ultraparallels.

ii. Any two lines of $\mathbf{H}^2$ determine a unique pencil.

We parametrize lines of $\mathbf{H}^2$ much as we did in $\mathbf{S}^2$. Let e_3 be an arbitrary point of $\mathbf{H}^2$. Let e_1 and e_2 be vectors of $\mathbf{R}^3$ such that $\{e_1, e_2, e_3\}$ is an orthonormal basis.

A typical point on the plane through the origin spanned by $\{e_1, e_3\}$ is $\lambda e_3 + \mu e_1$. This point is on $\mathbf{H}^2$ if and only if $\lambda > 0$ and

$$b(\lambda e_3 + \mu e_1, \lambda e_3 + \mu e_1) = -1;$$

that is,

$$\lambda^2 = 1 + \mu^2.$$

Using Theorem 3F, we may call $\lambda = \cosh t$ and $\mu = \sinh t$. Then as t ranges through all real numbers, $(\cosh t)e_3 + (\sinh t)e_1$ runs through all the points of the line. We define distance in such a way that t measures distance along the line.

Definition. *For x, y in $\mathbf{H}^2$ define*

$$d(x, y) = \cosh^{-1}(-b(x, y)).$$

Remark: This definition is possible because $b(x, y) \leqslant -1$, as was shown in the corollary to Theorem 4.

Theorem 9. *Let $\alpha(t) = (\cosh t)e_3 + (\sinh t)e_1$. Then*

$$d(\alpha(t_1), \alpha(t_2)) = |t_1 - t_2|.$$

Proof: Exercise 6. $\square$

Definition. *If $t_1 < t < t_2$, then $\alpha(t)$ is* between *$\alpha(t_1)$ and $\alpha(t_2)$.*

Now that we have defined distance between two points in the hyperbolic plane, it is necessary to determine which of the properties of Euclidean distance carry over to the hyperbolic case. The following is immediate from the definition.

Theorem 10. *If P and Q are points of $\mathbf{H}^2$, then*
  i. $d(P, Q) \geqslant 0$.
 ii. $d(P, Q) = 0$ *if and only if $P = Q$.*
iii. $d(P, Q) = d(Q, P)$.

We now address ourselves to the triangle inequality. Our proof of the triangle inequality in the spherical case relied on the cross product operation of $\mathbf{E}^3$. Here we use the hyperbolic cross product.

Theorem 11 (Triangle inequality). *Let P, Q, and R be points of* $\mathbf{H}^2$. *Then* $d(P, Q) + d(P, R) \geqslant d(Q, R)$ *with equality if and only if P, Q, R are collinear and P lies between Q and R.*

Proof: If P, Q, and R are not collinear, then $P \times Q$ and $R \times Q$ will not be proportional. Thus, $(P \times Q) \times (R \times Q) = b(P \times Q, R)Q$ is timelike. We may apply the hyperbolic Cauchy–Schwarz inequality (Theorem 3) to get

$$b(P \times Q, R \times Q)^2 \leqslant b(P \times Q, P \times Q)b(R \times Q, R \times Q). \quad (7.5)$$

But

$$
\begin{aligned}
b(P \times Q, R \times Q) &= b((P \times Q) \times R, Q) \\
&= -b(P, R)b(Q, Q) + b(Q, R)b(P, Q) \\
&= b(P, R) + b(Q, R)b(P, Q)
\end{aligned}
$$

because $b(Q, Q) = -1$. Let $d(Q, R) = p$, $d(P, R) = q$, $d(P, Q) = r$. Then

$$\cosh p = -b(Q, R), \quad \cosh r = -b(Q, P), \quad \cosh q = -b(P, R).$$

Thus,

$$b(P \times Q, R \times Q) = \cosh p \cosh r - \cosh q.$$

Also

$$
\begin{aligned}
b(P \times Q, P \times Q) &= -b(P, P)b(Q, Q) + b(R, Q)^2 \\
&= -1 + \cosh^2 r = \sinh^2 r.
\end{aligned}
$$

and, similarly, $b(R \times Q, R \times Q) = \sinh^2 p$. Equation (7.5) now becomes

$$(\cosh p \cosh r - \cosh q)^2 \leqslant \sinh^2 r \sinh^2 p.$$

Hence,

$$\cosh p \cosh r - \cosh q \leqslant \sinh r \sinh p,$$

$$\cosh q \geqslant \cosh(p - r),$$

$$q \geqslant p - r,$$

$$p \leqslant q + r.$$

This is what we wanted to prove. Now if $p = q + r$, we have equality in (7.5). From Theorem 3 this means that $(P \times Q) \times (R \times Q)$ is not timelike, and, hence, $b(P \times Q, R) = 0$; that is, R lies on $\overleftrightarrow{PQ}$. The fact that P lies between Q and R can be deduced easily from Theorem 9 and is left as an exercise (Exercise 7). □

Remark: The properties of the hyperbolic functions used in this section may be found in Appendix F.

Isometries of $\mathbf{H}^2$

A map $T: \mathbf{H}^2 \to \mathbf{H}^2$ is called an *isometry* if for all X and Y in $\mathbf{H}^2$,

$$d(TX, TY) = d(X, Y).$$

As in the case of $\mathbf{E}^2$, $\mathbf{S}^2$, and $\mathbf{P}^2$, isometries preserve collinearity. Specifically, we have the following.

Theorem 12. *Let T be an isometry of $\mathbf{H}^2$. Then three distinct points P, Q, and R of $\mathbf{H}^2$ are collinear if and only if TP, TQ, and TR are collinear.*

Proof: Exercise 8. □

Reflections

Let α be a line of $\mathbf{H}^2$ with unit normal ξ. For $x \in \mathbf{R}^3$ let

$$\Omega_\alpha x = x - 2b(x, \xi)\xi.$$

Theorem 13.
  i. $\Omega_\alpha^2 = I.$
 ii. Ω_α *is a bijection of $\mathbf{R}^3$ onto $\mathbf{R}^3$.*
iii. $b(\Omega_\alpha x, \Omega_\alpha Y) = b(x, y)$ *for all $x, y \in \mathbf{R}^3$.*

Proof:
  i. $\Omega_\alpha \Omega_\alpha x = \Omega_\alpha x - 2b(\Omega_\alpha x, \xi)\xi$
$\qquad\qquad = x - 2b(x, \xi)\xi - 2b(x, \xi)\xi + 4b(x, \xi)b(\xi, \xi)$
$\qquad\qquad = x.$
 ii. Follows easily from (i).
iii. $b(\Omega_\alpha x, \Omega_\alpha y) = b(x - 2b(x, \xi)\xi, y - 2b(y, \xi)\xi)$
$\qquad\qquad = b(x, y) - 2b(x, \xi)b(\xi, y) - 2b(y, \xi)b(x, \xi)$
$\qquad\qquad\quad + 4b(x, \xi)b(y, \xi)b(\xi, \xi)$
$\qquad\qquad = b(x, y).$ □

Corollary. *For any line α of $\mathbf{H}^2$ and $x \in \mathbf{R}^3$, we have the following:*
  i. *If x is timelike, so is $\Omega_\alpha x$.*
 ii. *If x is lightlike, so is $\Omega_\alpha x$.*
iii. *If x is spacelike, so is $\Omega_\alpha x$.*
 iv. *If x is a unit vector, so is $\Omega_\alpha x$.*
  v. *If $x \in \mathbf{H}^2$, so is $\Omega_\alpha x$.*

Definition. *Given a line α of $\mathbf{H}^2$, the restriction of Ω_α to $\mathbf{H}^2$ is called the* reflection *in α.*

Theorem 14. *Every reflection is an isometry of* $\mathbf{H}^2$.

Proof: For $X, Y \in \mathbf{H}^2$,

$$d(\Omega_\alpha X, \Omega_\alpha Y) = \cosh^{-1}(-b(\Omega_\alpha X, \Omega_\alpha Y))$$
$$= \cosh^{-1}(-b(X, Y)) = d(X, Y). \qquad \square$$

Theorem 15. *Let* β *be a line of* $\mathbf{H}^2$ *with unit normal* η. *Then*

$$\Omega_\alpha\beta = \{X \in \mathbf{H}^2 | b(X, \Omega_\alpha\eta) = 0\};$$

that is, if β *has unit normal* η, *then* $\Omega_\alpha\beta$ *is a line with unit normal* $\Omega_\alpha\eta$.

Proof: Let $Y \in \Omega_\alpha\beta$. Then for some $X \in \beta$, $Y = \Omega_\alpha X$ and

$$b(Y, \Omega_\alpha\eta) = b(\Omega_\alpha X, \Omega_\alpha\eta) = b(X, \eta) = 0.$$

Conversely, if $b(X, \Omega_\alpha\eta) = 0$, then $b(\Omega_\alpha X, \eta) = b(\Omega_\alpha\Omega_\alpha X, \Omega_\alpha\eta) = 0$. In other words, $\Omega_\alpha X \in \beta$ and $X \in \Omega_\alpha\beta$. $\qquad \square$

Theorem 16.
i. *Let* x *be a point of* $\mathbf{H}^2$. *Then* $\Omega_\alpha x = x$ *if and only if* $x \in \alpha$.
ii. *Let* β *be a line of* $\mathbf{H}^2$. *Then* $\Omega_\alpha\beta = \beta$ *if and only if* $\alpha = \beta$ *or* $\alpha \perp \beta$.

Proof:
i. $x - 2b(x, \xi)\xi = x$ if and only if $b(x, \xi) = 0$.
ii. Let $\beta = \{x \in \mathbf{H}^2 | b(x, \eta) = 0\}$, where $b(\eta, \eta) = 1$. Then $\Omega_\alpha\beta = \beta$ if and only if $\eta - 2b(\eta, \xi)\xi = \pm\eta$. This holds if and only if $b(\eta, \xi) = 0$ or $\xi = \pm\eta$. The former means that $\alpha \perp \beta$, and the latter means that $\alpha = \beta$. $\qquad \square$

Motions

As before, an isometry that is a product of reflections is called a *motion*. In addition to reflections we distinguish four special kinds of motions.

Let α and β be lines of $\mathbf{H}^2$. If α and β intersect in a point P of $\mathbf{H}^2$, then $\Omega_\alpha\Omega_\beta$ is called a *rotation about* P.

If α and β are parallel, then $\Omega_\alpha\Omega_\beta$ is called a *parallel displacement*. If α and β are ultraparallel with common perpendicular ℓ, then $\Omega_\alpha\Omega_\beta$ is called a *translation along* ℓ.

A *glide reflection* in $\mathbf{H}^2$ is the product of reflection in a line ℓ with a translation along ℓ. The line ℓ is called the *axis* of the glide reflection.

Rotations

Let P be an arbitrary point of $\mathbf{H}^2$. The set of rotations about P is denoted by ROT(P). We construct matrices representing each element of ROT(P)

and prove that ROT(P) is a group isomorphic to **SO(2)**.

Choose an orthonormal basis $\{e_1, e_2, e_3\}$ so that $e_3 = P$. If α is a line through P, we can write

$$\alpha = \{x \mid b(x, \xi) = 0\},$$

where

$$\xi = (-\sin\theta)e_1 + (\cos\theta)e_2.$$

Then

$$\Omega_\alpha e_1 = e_1 - 2(-\sin\theta)\xi = (\cos 2\theta)e_1 + (\sin 2\theta)e_2,$$

$$\Omega_\alpha e_2 = e_2 - (2\cos\theta)\xi = (\sin 2\theta)e_1 - (\cos 2\theta)e_1,$$

$$\Omega_\alpha e_3 = e_3.$$

Thus, the matrix of Ω_α with respect to $\{e_1, e_2, e_3\}$ is

$$\begin{bmatrix} \cos 2\theta & \sin 2\theta & 0 \\ \sin 2\theta & -\cos 2\theta & 0 \\ 0 & 0 & 1 \end{bmatrix} = \begin{bmatrix} \text{ref } \theta & & 0 \\ & & 0 \\ 0 & 0 & 1 \end{bmatrix}.$$

Now, if β is another line through P with pole

$$\eta = (-\sin\phi)e_1 + (\cos\phi)e_2,$$

then $\Omega_\alpha\Omega_\beta$ takes a similar form with θ replaced by ϕ. By the calculations of Chapter 1, the matrix of $\Omega_\alpha\Omega_\beta$ is

$$\begin{bmatrix} \cos 2(\theta-\phi) & -\sin 2(\theta-\phi) & 0 \\ \sin 2(\theta-\phi) & \cos 2(\theta-\phi) & 0 \\ 0 & 0 & 1 \end{bmatrix} = \begin{bmatrix} \text{rot } 2(\theta-\phi) & & 0 \\ & & 0 \\ 0 & 0 & 1 \end{bmatrix}.$$

The function $\Omega_\alpha \to$ ref θ determines an isomorphism of REF(P) (the group generated by reflections of $\mathbf{H}^2$ in lines through P) onto **O(2)**. Under this isomorphism ROT(P) goes into **SO(2)**. Recalling the formulas of Chapter 1 (Theorems 33 and 34), we conclude the following:

Theorem 17 (Three reflections theorem). *Let α, β, and γ be lines through P in $\mathbf{H}^2$. Then there is a fourth line δ through P such that*

$$\Omega_\alpha\Omega_\beta\Omega_\gamma = \Omega_\delta.$$

The related representation theorem for rotations holds.

Theorem 18. *Let ρ be a rotation about P. Let ℓ be a line through P. Then there exist lines m and m' through P such that*

$$\rho = \Omega_\ell\Omega_m = \Omega_{m'}\Omega_\ell.$$

$\mathbf{H}^2$ as a subset of $\mathbf{P}^2$

As well as being an interesting subject of study in its own right, the projective plane provides a framework in which other geometries can be embedded, often allowing an approach that facilitates both computation and understanding. In Chapter 5 we saw that it was possible to regard the incidence geometry of $\mathbf{E}^2$ as a subgeometry of $\mathbf{P}^2$.

We now show that the hyperbolic plane can also be regarded as a subgeometry of $\mathbf{P}^2$. Let $\mathbf{D}^2$ be the subset of $\mathbf{P}^2$ determined by the condition $b(x, x) < 0$. This set of points may be regarded as the interior of the conic $b(x, x) = 0$. We will call the remaining points of $\mathbf{P}^2$ (those with $b(x, x) > 0$) *exterior* points.

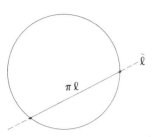

Figure 7.3 The Klein model. $\tilde{\ell} \cap \mathbf{D}^2$ represents a line of the hyperbolic plane.

Theorem 19.
   i.   *The usual projection* $\pi\colon \mathbf{R}^3 - \{0\} \to \mathbf{P}^2$ *maps* $\mathbf{H}^2$ *bijectively to* $\mathbf{D}^2$.
   ii. *For each point X of $\mathbf{P}^2$ exterior to $\mathbf{D}^2$, there is a unique pair $\{\xi, -\xi\}$ of unit spacelike vectors such that $\pi\xi = \pi(-\xi) = X$. Conversely, each unit spacelike vector determines such an exterior point.*
   iii. *A vector $v \in \mathbf{R}^3$ is lightlike if and only if πv lies on the conic.*

For most purposes we can look at $\mathbf{D}^2$ as the unit disk $x_1^2 + x_2^2 < 1$ in the plane $x_3 = 1$ of $\mathbf{E}^3$ and work in this model of $\mathbf{E}^2$ rather than in $\mathbf{P}^2$. We use the correspondence defined in Chapter 5, which relates $\mathbf{E}^2$ and $\mathbf{P}^2 - \ell_\infty$. In terms of homogeneous coordinates, ℓ_∞ is the line $x_3 = 0$.

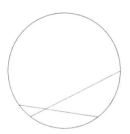

Figure 7.4 Intersecting lines.

Theorem 20.   *In terms of the model described in Theorem 19, if ℓ is a line of $\mathbf{H}^2$ then $\pi\ell$ is a chord of the disk $\mathbf{D}^2$. The end points of the chord are, of course, not included in $\mathbf{D}^2$ nor in $\pi\ell$. (See Figure 7.3.)*

Remark: If ℓ is a line of $\mathbf{H}^2$, then $\pi\ell$ is contained in a unique line $\tilde{\ell}$ of $\mathbf{P}^2$. On the other hand, not all lines of $\mathbf{P}^2$ determine lines of $\mathbf{H}^2$; only those that are secants of the conic.

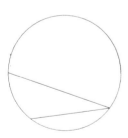

Figure 7.5 Parallel lines.

Theorem 21.   *Let ℓ_1 and ℓ_2 be lines of $\mathbf{H}^2$. Then*
   i.   *ℓ_1 and ℓ_2 are intersecting lines if and only if $\tilde{\ell}_1$ and $\tilde{\ell}_2$ intersect in $\mathbf{D}^2$. (See Figure 7.4.)*
   ii. *ℓ_1 and ℓ_2 are parallel if and only if $\tilde{\ell}_1$ and $\tilde{\ell}_2$ intersect at a point on the boundary of $\mathbf{D}^2$. (See Figure 7.5.)*
   iii. *ℓ_1 and ℓ_2 are ultraparallel if and only if $\tilde{\ell}_1$ and $\tilde{\ell}_2$ intersect at a point exterior to $\mathbf{D}^2$. (See Figure 7.6.)*

Theorem 22.   *Two lines ℓ_1 and ℓ_2 are perpendicular if and only if $\tilde{\ell}_1$ and $\tilde{\ell}_2$ are conjugate.*

Theorem 23. *Let ℓ be a line of $\mathbf{H}^2$. Let P and Q be points of $\mathbf{P}^2$ where $\tilde{\ell}$ meets the conic. Then the pole of $\tilde{\ell}$ is the intersection R of the respective tangents through P and Q.*

Corollary. *A line m of $\mathbf{H}^2$ is perpendicular to ℓ if and only if $\tilde{m}$ passes through R, the pole of $\tilde{\ell}$. Figure 7.7 illustrates this and the previous two theorems.*

Theorem 24. *Each point of $\mathbf{P}^2$ determines a unique pencil of $\mathbf{H}^2$ as follows:*
  i. *Each point $\pi x = P$ of $\mathbf{D}^2$ determines the pencil of intersecting lines through $x \in \mathbf{H}^2$.*
 ii. *Each point $\pi v = P$ (where v is lightlike) determines a pencil of parallels.*
iii. *Each point $\pi\xi = P$ (where ξ is a unit spacelike vector) determines a pencil of ultraparallels. The common perpendicular to this pencil corresponds to the polar line of P.*

In each case the pencil consists of all lines ℓ of $\mathbf{H}^2$ such that $\tilde{\ell}$ passes through the designated point P of $\mathbf{P}^2$.

Remark: The pictures in Figures 7.8–7.10 give an intuitive idea of these relationships.

The discussion of this section should provide a motivation for some of the constructions we have been making in hyperbolic geometry. The incidence geometry of $\mathbf{D}^2$ is precisely that of $\mathbf{H}^2$, and this model of $\mathbf{H}^2$ is called the *Klein model*. Unfortunately, the Klein model does not represent either distance or angle faithfully, so it is unwise to rely too heavily on it. For example, a line is infinitely long, although it is represented in $\mathbf{D}^2$ by a (finite) chord.

Parallel displacements

Let $\mathscr{P}$ be a pencil of parallels determined by two lines with unit normals ξ and η. Choose an orthonormal basis by setting $e_1 = \xi$, $e_3 \in \mathbf{H}^2$, and $e_2 = e_3 \times e_1$. If we write

$$\eta = \lambda e_1 + \mu e_2 + v e_3, \quad \text{with} \quad \lambda \geq 0,$$

the conditions that $b(\eta, \eta) = 1$ and that $\xi \times \eta$ is lightlike give $\mu = \pm v$ and $\lambda = 1$. Hence,

$$\eta = e_1 + \mu(e_2 \pm e_3).$$

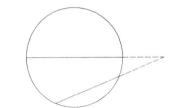

Figure 7.6 Ultraparallel lines.

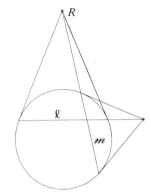

Figure 7.7 Two perpendicular lines, ℓ and m.

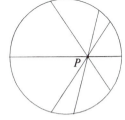

Figure 7.8 A pencil of intersecting lines.

163

Theorem 25. *In terms of the basis just described, suppose that $\mathcal{P}$ contains lines with unit normals $(1, 0, 0)$ and $(1, \mu, -\mu)$ for some real number μ. Then $\mathcal{P}$ consists precisely of those lines with unit normals of the form $(1, r, -r)$, where r ranges through the real numbers.*

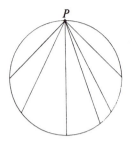

Figure 7.9 A pencil of parallels.

Proof: First note that

$$\xi \times \eta = \mu(e_1 \times e_2 - e_1 \times e_3) = -\mu(e_3 - e_2),$$

so that $\xi \times \eta$ is lightlike. Furthermore, if $\zeta = e_1 + r(e_2 - e_3)$, then

$$b(\zeta, \xi \times \eta) = \mu r b(e_2 - e_3, e_2 - e_3) = 0.$$

Conversely, if ζ is a unit spacelike vector orthogonal to $\xi \times \eta$, it is easy to check that $\pm\zeta$ must be of the form $(1, r, -r)$. $\square$

Remark: The projective model of $\mathbf{H}^2$ provides some insight into what is going on here. The self-conjugate point P of $\mathbf{P}^2$ through which all lines of the pencil pass is $(0, 1, -1)$. A typical line of the pencil has its pole on the tangent to the conic at P. (See Figure 7.11.)

Note that a line of $\mathbf{H}^2$ belongs to two distinct pencils. In our example there is a second pencil through $Q = (0, 1, 1)$. The same basis may be used, but in this case the unit normals of lines of the pencil are $(1, r, r)$.

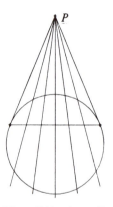

Figure 7.10 A pencil of ultraparallels.

We have shown that a pencil of parallels is parametrized by the set of real numbers. Let α be a line of the pencil with pole $(1, r, -r)$ in homogeneous coordinates. Then

$$\Omega_\alpha e_1 = e_1 - 2b(\xi, e_1)\xi = -e_1 - 2re_2 + 2re_3,$$

$$\Omega_\alpha e_2 = e_2 - 2r(e_1 + re_2 - re_3) = -2re_1 + (1 - 2r^2)e_2 + 2r^2e_3,$$

$$\Omega_\alpha e_3 = e_3 - 2r(e_1 + re_2 - re_3) = -2re_1 - 2r^2e_2 + (1 + 2r^2)e_3.$$

The matrix of Ω_α is

$$\begin{bmatrix} -1 & -2r & -2r \\ -2r & 1 - 2r^2 & -2r^2 \\ 2r & 2r^2 & 1 + 2r^2 \end{bmatrix}. \tag{7.6}$$

If β is a second line of this pencil, a calculation shows that

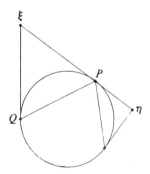

Figure 7.11 Two lines of a parallel pencil.

$$\Omega_\alpha \Omega_\beta = \begin{bmatrix} 1 & 2h & 2h \\ -2h & 1 - 2h^2 & -2h^2 \\ 2h & 2h^2 & 1 + 2h^2 \end{bmatrix} = D_h, \tag{7.7}$$

where β has pole $(1, s, -s)$ and $h = s - r$. Thus, the parallel displacement $\Omega_\alpha \Omega_\beta$ is represented with respect to this basis by the matrix D_h of (7.7). One can check that for real numbers h and k,

$$D_h D_k = D_{h+k}.$$

Theorem 26 (Three reflections theorem). *Let α, β, and γ be lines in a pencil of parallels. Then there is a fourth line δ in the pencil such that*

$$\Omega_\alpha \Omega_\beta \Omega_\gamma = \Omega_\delta.$$

Proof: With respect to an appropriate basis of $\mathbf{R}^3$, there exist real numbers r, s, and t representing α, β, and γ in the sense that $(1, r, -r)$ is a unit normal to α, and so forth. Now

$$\Omega_\alpha \Omega_\beta \Omega_\gamma = \dot\Omega_\delta \quad \text{iff} \quad \Omega_\beta \Omega_\gamma = \Omega_\alpha \Omega_\delta;$$

that is,

$$D_{t-s} = D_{u-r},$$

where u is the real number representing δ. If we choose $u = r + t - s$, this last equation becomes true. Hence, the theorem is true, and the pole of the required line δ is represented by

$$e_1 + (r + t - s)(e_2 - e_3). \qquad \Box$$

A representation theorem for parallel displacements holds also.

Theorem 27. *Let ρ be a parallel displacement arising from a pencil $\mathscr{P}$. Let ℓ be a line of $\mathscr{P}$. Then there are lines m and m' in $\mathscr{P}$ such that*

$$\rho = \Omega_\ell \Omega_m = \Omega_{m'} \Omega_\ell.$$

Proof: Exercise 18. $\qquad \Box$

Let $\mathrm{REF}(\mathscr{P})$ be the group generated by all reflections in lines of the pencil $\mathscr{P}$. Let $\mathrm{DIS}(\mathscr{P})$ be the set of all parallel displacements determined by the pencil $\mathscr{P}$. In Exercise 19 you will show that $\mathrm{DIS}(\mathscr{P})$ is a group and investigate its algebraic properties.

Translations

Let $\mathscr{P}$ be an ultraparallel pencil with common perpendicular ℓ. Let e_1 be a unit normal of ℓ. Choose e_2 and e_3 spacelike and timelike, respectively, so that $\{e_1, e_2, e_3\}$ is an orthonormal basis.

Let α be an arbitrary line of the pencil. Its unit normal can be written

$$\xi = (\cosh u)e_2 + (\sinh u)e_3.$$

Then

$$\Omega_\alpha e_1 = e_1 - 2b(e_1, \xi)\xi = e_1,$$

$$\Omega_\alpha e_2 = e_2 - (2 \cosh u)((\cosh u)e_2 + (\sinh u)e_3)$$
$$= -(\cosh 2u)e_2 - (\sinh 2u)e_3,$$

$$\Omega_\alpha e_3 = (\sinh 2u)e_2 + (\cosh 2u)e_3.$$

$$\Omega_\alpha = \begin{bmatrix} 1 & 0 & 0 \\ 0 & -\cosh 2u & \sinh 2u \\ 0 & -\sinh 2u & \cosh 2u \end{bmatrix}. \tag{7.8}$$

If β is a second line of the pencil whose pole is parametrized by v, then

$$\Omega_\alpha \Omega_\beta = \begin{bmatrix} 1 & 0 & 0 \\ 0 & \cosh 2k & \sinh 2k \\ 0 & \sinh 2k & \cosh 2k \end{bmatrix}, \tag{7.9}$$

where $k = u - v$.

Denote this last matrix by T_k. Then one can easily verify that

$$T_k T_m = T_{k+m}.$$

Theorem 28 (Three reflections theorem). *Let α, β, and γ be lines of a pencil of ultraparallels. Then there is a line δ in the pencil such that*

$$\Omega_\alpha \Omega_\beta \Omega_\gamma = \Omega_\delta.$$

Proof: Exercise 21. ☐

Theorem 29 (Representation of translations). *Let ρ be a translation along a line ℓ. Let m be any line perpendicular to ℓ. Then there exist lines α and α' perpendicular to ℓ such that*

$$\rho = \Omega_m \Omega_\alpha = \Omega_{\alpha'} \Omega_m.$$

Let $\mathcal{P}$ be a pencil of ultraparallels, and let ℓ be the common perpendicular. The group generated by all reflections in lines of $\mathcal{P}$ is denoted by REF($\mathcal{P}$). Let TRANS(ℓ) be the set of translations along ℓ. Properties of TRANS(ℓ) will be left to the exercises (Exercise 22).

Glide reflections

With respect to the basis used in the previous section, we construct the matrix of the glide reflection $\Omega_\ell T_k$. One can easily check that

$$\Omega_\ell e_1 = -e_1, \quad \Omega_\ell e_2 = e_2, \quad \Omega_\ell e_3 = e_3.$$

Thus,

$$\Omega_\ell T_k = \begin{bmatrix} -1 & 0 & 0 \\ 0 & \cosh 2k & \sinh 2k \\ 0 & \sinh 2k & \cosh 2k \end{bmatrix}. \tag{7.10}$$

Products of more than three reflections

In each of the geometries studied so far, any motion can be realized as the product of two or three reflections. The same is true in $\mathbf{H}^2$. However, the incidence structure of $\mathbf{H}^2$ is more complicated. More cases must be considered in the proof.

Our approach is to show that any product $\Omega_\alpha \Omega_\beta \Omega_\gamma \Omega_\delta$ of four reflections can be reduced to a product of two reflections (as in Theorem 1.36). As a first observation, if the pencil determined by α and β has a line in common with the pencil determined by γ and δ, our representation theorems may be applied to rewrite our product of four reflections in such a way that the second and third reflections are the same.

We begin with

Theorem 30. *Let P be a point of $\mathbf{H}^2$, and let $\mathscr{P}$ be a pencil. Then there is a line through P belonging to the pencil $\mathscr{P}$. Except in the case of the pencil of all lines through P, this line is unique.*

Proof: If $\mathscr{P}$ is a pencil of intersecting lines or a pencil of ultraparallels, the conclusion is given by Theorem 4 and Theorem 8, part (ii), respectively. Now let $\mathscr{P}$ be a pencil of parallels determined by lines with unit normal ξ and η. Then $\xi \times \eta$ is lightlike, and $P \times (\xi \times \eta)$ is nonzero. By Theorem 2, part (ii), $P \times (\xi \times \eta)$ is spacelike. The line whose unit normal is in this direction belongs to $\mathscr{P}$ and passes through P and is the only line satisfying these conditions. $\qquad\square$

Theorem 31. *Let $\mathscr{P}_1$ be a pencil of parallels. Let $\mathscr{P}_2$ be the pencil consisting of all lines perpendicular to a line γ. If $\gamma \notin \mathscr{P}_1$, there is a unique line belonging to both pencils.*

Proof: Choose an orthonormal basis as follows. Let e_2 be a unit normal to γ. Let w be a lightlike vector such that the unit normals ξ to lines of $\mathscr{P}_1$ are precisely those unit spacelike vectors satisfying $b(\xi, w) = 0$. See Figure 7.12.

We wish to choose $e_3 \in \mathbf{H}^2$ so that it lies in $[w, e_2]$. To do this, note that

$$b(w \times e_2, w \times e_2) = b(w, e_2)^2 > 0$$

because $\gamma \notin \mathscr{P}_1$. Choose e_1 to be a unit vector in the direction $[w \times e_2]$, and $e_3 = e_1 \times e_2$. There are two choices for e_1, but only one that will ensure that e_3 lies in $\mathbf{H}^2$.

Now that we have this basis, it is easy to see that the line with unit normal e_1 is the unique line belonging to both pencils. $\qquad\square$

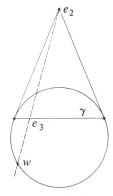

Figure 7.12 Construction of a line common to a parallel pencil and an ultraparallel pencil.

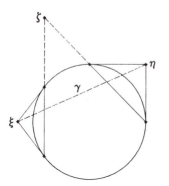

Figure 7.13 Theorem 7. Two ultra-parallel pencils with a common line γ.

Remark: If the two pencils are related in such a way that $\gamma \in \mathscr{P}_1$, then they can have no line in common. (See Exercise 33.)

Theorem 32. *Two distinct pencils of parallels have a unique line in common.*

Proof: Let v and w be lightlike vectors determining distinct pencils. Then Theorem 5.26 gives

$$b(v \times w, v \times w) = b(v, w)^2,$$

which is positive (Exercise 34). The line whose unit normal is a multiple of $v \times w$ is the unique line common to both pencils. □

Remark: Two ultraparallel pencils have a line in common if and only if the common perpendiculars to the two pencils are themselves ultraparallel. See Figure 7.13 and Theorem 7. Thus, we have completed our analysis of the question of when two pencils have a line in common. We now have enough ammunition to attempt the task set out at the beginning of this section.

Theorem 33. *Let α, β, γ, and δ be lines. Then there exist lines u and v such that*

$$\Omega_\alpha \Omega_\beta \Omega_\gamma \Omega_\delta = \Omega_u \Omega_v.$$

Proof: If $\alpha = \beta$ or $\gamma = \delta$, there is nothing to prove. Assume that α and β determine a pencil $\mathscr{P}_1$, whereas γ and δ determine a pencil $\mathscr{P}_2$. As remarked at the beginning of this section, the result clearly holds if $\mathscr{P}_1$ and $\mathscr{P}_2$ have a line in common. In view of Theorems 30–32, we have still to consider the following cases:

1. α and β have γ as a common perpendicular, and δ is parallel to γ. In this case Ω_β commutes with Ω_γ, and Theorem 30 applies.

2. α and β have a common perpendicular ℓ, γ and δ have a common perpendicular m, and ℓ intersects m in a point P. Using Theorem 29, we may replace the given representation by $\Omega_{\alpha'} \Omega_{\beta'} \Omega_{\gamma'} \Omega_{\delta'}$, where β' and γ' pass through P. Then $\Omega_{\beta'} \Omega_{\gamma'}$ may be replaced by $\Omega_\ell \Omega_n$ for some line n through P. Because α' is perpendicular to ℓ, Theorem 30 now applies.

3. α and β have a common perpendicular ℓ, γ and δ have a common perpendicular m, but ℓ is parallel to m. In this case we let Q be the point where m intersects γ, and we let β' be the line through Q perpendicular to ℓ. Then the motion can be written $\Omega_{\alpha'} \Omega_{\beta'} \Omega_\gamma \Omega_\delta$ for some line $\alpha' \perp \ell$. As in case (2) we may now write

$$\Omega_{\beta'} \Omega_\gamma = \Omega_n \Omega_m$$

for some line n. Again apply Theorem 30. □

Remark: Because

$$(\Omega_\alpha\Omega_\beta\Omega_\gamma\Omega_\delta)^{-1} = \Omega_\delta\Omega_\gamma\Omega_\beta\Omega_\alpha,$$

the foregoing set of cases is exhaustive. For example, it is not necessary to consider the case where α and β are parallel while γ and δ have a common perpendicular.

Theorem 34. *Let α, β, and γ be lines not belonging to any pencil. Then $\Omega_\alpha\Omega_\beta\Omega_\gamma$ is a nontrivial glide reflection.*

The proof of Theorem 34 uses techniques similar to those we have been using in Theorem 33. It is left as an exercise (Exercise 35).

We can now assert the following classification of motions of $\mathbf{H}^2$.

Theorem 35. *The group of motions of $\mathbf{H}^2$ consists of all reflections, rotations, translations, parallel displacements, and glide reflections. Every motion is the product of two or three suitably chosen reflections.*

Fixed points of isometries

Consider the isometry $\rho = \Omega_\alpha\Omega_\beta$. Fixed points of ρ are found by solving for $X \in \mathbf{H}^2$ the equation $\rho X = X$; that is, $\Omega_\alpha X = \Omega_\beta X$. Any solution must satisfy $b(X, \xi)\xi = b(X, \eta)\eta$. If $b(X, \xi) = b(X, \eta) = 0$, then X is a multiple of $\xi \times \eta$. This means that X is the point of intersection of α and β in $\mathbf{H}^2$. On the other hand, if $b(X, \xi) \neq 0$ or $b(X, \eta) \neq 0$, ξ must be a multiple of η, and so $\alpha = \beta$. Thus, we can state

Theorem 36.
  i. *A nontrivial translation has no fixed points.*
 ii. *A nontrivial rotation has exactly one fixed point, the center of rotation.*
iii. *A nontrivial parallel displacement has no fixed points.*
 iv. *A reflection has a line of fixed points, the axis of reflection.*
  v. *A nontrivial glide reflection has no fixed points.*

This result may be compared with the Euclidean analogue, Theorem 1.39. For the proof see Exercise 37.

Fixed lines of isometries

If Ω_α is a reflection whose axis has unit normal ξ, then Ω_α will leave fixed the lines whose unit normals ζ satisfy $\Omega_\alpha\zeta = \pm\zeta$; that is, ζ must be orthogonal to ξ or $\zeta = \pm\xi$.

Suppose now that α and β are lines with respective unit normals ξ

and η. Then $\rho = \Omega_\alpha \Omega_\beta$ has a fixed line with unit normal ξ if and only if $\Omega_\alpha \zeta = \pm \Omega_\beta \zeta$. The condition $\Omega_\alpha \zeta = -\Omega_\beta \zeta$ is satisfied only if $\zeta = b(\zeta, \xi)\xi + b(\zeta, \eta)\eta$. But this implies that

$$b(\zeta, \xi) = b(\zeta, \xi)b(\xi, \xi) + b(\zeta, \eta)b(\eta, \xi)$$

and

$$b(\zeta, \eta) = b(\zeta, \xi)b(\zeta, \eta) + b(\zeta, \eta)b(\eta, \eta).$$

These equations in turn give

$$b(\zeta, \eta)b(\eta, \xi) = b(\zeta, \xi)b(\eta, \xi) = 0.$$

Thus, either $\alpha \perp \beta$ or $b(\zeta, \eta) = b(\zeta, \xi) = 0$, which would imply $\zeta = 0$, an impossibility.

We conclude that $\Omega_\alpha \zeta = -\Omega_\beta \zeta$ if and only if $\alpha \perp \beta$ and ζ is in the span of ξ and η. This means that ρ is a half-turn, and the fixed line passes through its center.

We now search for ζ with $b(\zeta, \zeta) = 1$ and $\Omega_\alpha \zeta = \Omega_\beta \zeta$. Then $b(\zeta, \xi)\xi = b(\zeta, \eta)\eta$. If $\alpha \neq \beta$, this implies that $b(\zeta, \xi) = b(\zeta, \eta) = 0$. Thus, the line of $\mathbf{H}^2$ with unit normal ζ is a common perpendicular of α and β.

Summarizing our results concerning isometries that are the product of two reflections, we have

Theorem 37. *Let α and β be distinct lines of $\mathbf{H}^2$. The isometry $\Omega_\alpha \Omega_\beta$ has the following fixed line behavior.*
   i. *If α and β intersect at P and $\alpha \perp \beta$, every line through P is fixed. In this case, $\Omega_\alpha \Omega_\beta$ is the half-turn about P.*
   ii. *If α and β intersect at P and α is not perpendicular to β, $\Omega_\alpha \Omega_\beta$ has no fixed lines.*
   iii. *If α and β are parallel, $\Omega_\alpha \Omega_\beta$ has no fixed lines. Thus, parallel displacements have no fixed lines.*
   iv. *If α and β have a common perpendicular ℓ, then $\Omega_\alpha \Omega_\beta$ leaves ℓ fixed but has no other fixed lines.*

Theorem 38. *Let γ be a line perpendicular to two distinct lines α and β. Then the nontrivial glide reflection $\Omega_\alpha \Omega_\beta \Omega_\gamma$ has γ as its only fixed line.*

Proof: We must determine the unit spacelike vectors ζ satisfying

$$\Omega_\alpha \zeta = \pm \Omega_\beta \Omega_\gamma \zeta.$$

A calculation similar to that used in Theorem 37 shows that the positive sign cannot occur. With the negative sign ζ must be a unit normal to the line γ. The details are left to Exercise 38. $\qquad \square$

Segments, rays, angles, and triangles

Let P be a point. As we know, a line through P can be parametrized by

$$\alpha(t) = (\cosh t)P + (\sinh t)\xi$$

for a suitable unit spacelike vector ξ. A set of the form $\alpha([0, L])$, $L > 0$, is called a *segment* of length L. The points $\alpha(0)$ and $\alpha(L)$ are called *end points*. The point $M = \alpha(L/2)$ is the *midpoint*, and the usual definition of perpendicular bisector holds. The set $\alpha([0, \infty))$ is called a *ray*. The point $\alpha(0)$ is called the *origin* of the ray. It is a not-quite-obvious fact that these definitions have all the properties we should expect.

Theorem 39.
i. *Two distinct points A and B are the end points of exactly one segment, which we denote by AB or, equivalently, BA. The length of AB is $d(A, B)$.*
ii. *Each ray has exactly one origin. For each pair of points A and B, there is exactly one ray with origin A that passes through B. We denote this ray by $\overrightarrow{AB}$.*

Definition. *The unit spacelike vector ξ occurring in the definition of α is called the* direction vector *of the ray $\alpha([0, \infty))$. Note that $b(P, \xi) = 0$.*

Remark: Each ray has a unique direction vector. Taking our inspiration from the projective model of $\mathbf{H}^2$, we may think of ξ as a point "past infinity" toward which the ray is heading.

Angles and triangles are defined as in $\mathbf{E}^2$ along with the associated terms (straight angles, opposite rays, etc.). The radian measure of an angle is $\cos^{-1} b(\xi, \eta)$, where ξ and η are the direction vectors of the rays making up the angle. In Exercise 41 you will be asked to check that this is equivalent to

$$\cos^{-1} b\left(\frac{Q \times P}{|Q \times P|}, \frac{Q \times R}{|Q \times R|}\right)$$

for the angle $\angle PQR$. Note that $Q \times P$ and $Q \times R$ are spacelike vectors.

Definition. *A half-plane* bounded by a line ℓ *is a set of the form*

$$\{x \in \mathbf{H}^2 \mid b(\xi, x) > 0\},$$

where ξ is a unit normal of ℓ.

Theorem 40. *Each half-plane is bounded by a unique line. Each line bounds two half-planes. The union of these two half-planes is $\mathbf{H}^2 - \ell$. Two*

171

points of $\mathbf{H}^2 - \ell$ *are in the same half-plane if and only if the segment joining them does not meet* ℓ.

Definition. *The* interior *of an angle* $\angle PQR$ *is the intersection of the half-plane bounded by* $\overleftrightarrow{PQ}$ *containing R with the half-plane bounded by* $\overleftrightarrow{RQ}$ *containing P. (This definition does not make sense for straight angles or zero angles. The interior of such an angle is undefined.)*

Theorem 41. *Let* $\mathscr{A} = \angle PQR$ *be an angle whose interior is defined. Let* ξ *and* η *be direction vectors of its arms. Then the interior of* $\mathscr{A}$ *consists of those points X such that the direction vector of* $\overrightarrow{QX}$ *is a positive linear combination of* ξ *and* η.

Proof: Let X be any point other than Q, and let ζ be the direction vector of $\overrightarrow{QX}$. Because the subspace $\{v \in \mathbf{R}^3 | b(v, Q) = 0\}$ is two dimensional, $\{\xi, \eta\}$ is a basis, and there are unique numbers λ and μ such that

$$\zeta = \lambda\xi + \mu\eta.$$

We claim that X is in the interior of $\mathscr{A}$ if and only if λ and μ are positive. To see this, write

$$X = (\cosh t)Q + (\sinh t)\zeta$$

and note that $\xi \times Q$ is a unit normal to one arm, say $\overrightarrow{QP}$. Then

$$b(X, \xi \times Q) = (\sinh t)b(\zeta, \xi \times Q) = \mu(\sinh t)b(\eta, \xi \times Q)$$

$$= \mu\frac{\sinh t}{\sinh s}b(R, \xi \times Q),$$

where s is the number satisfying $R = (\cosh s)Q + (\sinh s)\eta$. Thus, X and R lie on the same side of $\overleftrightarrow{PQ}$ if and only if $\mu > 0$. Similarly, X and P lie on the same side of $\overleftrightarrow{RQ}$ if and only if $\lambda > 0$. $\square$

Addition of angles

Theorem 42.

i. *Let* $\mathscr{A} = \angle PQR$ *be an angle with a point X in its interior. Then the radian measure of* $\mathscr{A}$ *is the sum of the radian measures of* $\angle PQX$ *and* $\angle RQX$.

ii. *Let* $\mathscr{A} = \angle PQR$ *be a straight angle, and let X be any point not on the line* $\overleftrightarrow{PQ}$. *Then the sum of the radian measures of* $\angle PQX$ *and* $\angle RQX$ *is equal to* π.

Proof:

i. Let ξ, η, and ζ be the respective direction vectors as in Theorem 41. We need to prove the identity

$$\cos^{-1} b(\xi, \zeta) + \cos^{-1} b(\zeta, \eta) = \cos^{-1} b(\xi, \eta), \qquad (7.11)$$

using the fact that $\zeta = \lambda\xi + \mu\eta$, where λ and μ are positive numbers satisfying

$$b(\zeta, \zeta) = \lambda^2 + \mu^2 + 2\lambda\mu b(\xi, \eta) = 1.$$

For convenience write $a = b(\xi, \eta)$. Then $b(\xi, \zeta) = \lambda + \mu a$ and $b(\zeta, \eta) = \lambda a + \mu$, so that our identity reduces to

$$\cos^{-1}(\lambda + \mu a) + \cos^{-1}(\lambda a + \mu) = \cos^{-1} a, \qquad (7.12)$$

which can be verified by calculus (Exercise 44).

ii. When $\angle PQR$ is a straight angle, we have no expression for ζ in terms of ξ and η. We do not need one, however. The required identity reduces to

$$\cos^{-1} b(\xi, \zeta) + \cos^{-1} b(-\xi, \zeta) = \pi,$$

which is just one of the standard properties of the $\cos^{-1}$ function. (See Theorem 2F.) $\qquad\square$

Remark: The Euclidean version of this theorem (Theorem 2.14) is essentially the same thing. We could have used the preceding proof in Chapter 2. On the other hand, if we fixed an orthonormal basis $\{e_1, e_2, e_3\}$ for $\mathbf{R}^3$ with $e_3 = Q$, the proof given in 2.34 can be easily modified to prove Theorem 42. Both proofs have advantages and disadvantages. The first one is more geometric. The second one is more direct but relies explicitly on a computation involving differentiation, and therefore it is in some sense less elementary.

Remark: All our Euclidean definitions of rectilinear figures and their associated properties hold true in $\mathbf{H}^2$. Because of the incidence structure of $\mathbf{H}^2$, however, some new types of figures are possible.

Triangles and hyperbolic trigonometry

In hyperbolic geometry triangles are easier to deal with than in spherical or elliptic geometry because segments are simple. Each pair of points determines a unique segment. Thus, we can define, as in $\mathbf{E}^2$, the triangle $\triangle PQR$ to be the union of the segments PQ, QR, and PR. Each triangle has three angles. The interior of the triangle is the intersection of the interiors of its three angles.

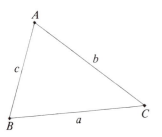

Figure 7.14 Theorem 43. A hyperbolic triangle.

Theorem 43. *Let ABC be a triangle. Let a, b, and c be the lengths of its sides. Then, using the same notational conventions as in spherical trigonometry (see Figure 7.14), we have*

i.
$$\cos A = \frac{\cosh b \, \cosh c - \cosh a}{\sinh b \, \sinh c}. \tag{7.13}$$

ii.
$$\frac{\sin A}{\sinh a} = \frac{2(\sinh s \, \sinh(s-a) \, \sinh(s-b) \, \sinh(s-c))^{1/2}}{\sinh a \, \sinh b \, \sinh c}. \tag{7.14}$$

iii.
$$\cosh a = \frac{\cos A + \cos B \, \cos C}{\sin B \, \sin C}. \tag{7.15}$$

Remark: Birman and Nomizu [5] have worked out trigonometric formulas for Lorentzian plane geometry. Their formulas bear a relationship to (7.13) and (7.14) analogous to that between plane Euclidean and spherical (Theorem 4.38) formulas, This is related to the fact that $\mathbf{H}^2$ may be regarded as a "sphere" in Lorentzian three-space. (See Exercise 72.)

Asymptotic triangles

Each line belongs to two parallel pencils. However, each ray determines a unique parallel pencil. In fact, if ζ is the direction vector of a ray $\overrightarrow{PX}$, then $P + \zeta$ is a lightlike vector with the property that $\{\xi | b(\xi, P + \zeta) = 0\}$ is the set of unit normal vectors of a unique pencil. (The other pencil to which the line $\overleftrightarrow{PX}$ belongs is determined by $P - \zeta$.)

Let PQ be a segment, and let $\overrightarrow{PX}$ and $\overrightarrow{QY}$ be parallel rays determining the same pencil. Then the union of PQ and the two rays is called *a (singly) asymptotic triangle*. Two views of an asymptotic triangle are shown in Figure 7.15. You may think of an asymptotic triangle as an ordinary triangle with one vertex "at ∞."

A pair of rays $\overrightarrow{PX}$ and $\overrightarrow{PY}$ together with the line common to the parallel pencils they determine is a *doubly asymptotic triangle*. See Figure 7.16. A *triply asymptotic triangle* consists of three lines mutually parallel in pairs. See Figure 7.17.

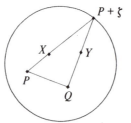 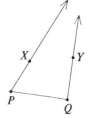

Figure 7.15 A singly asymptotic triangle, two views.

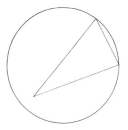

Figure 7.16   A doubly asymptotic triangle, two views.

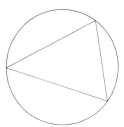

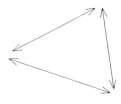

Figure 7.17   A triply asymptotic triangle, two views.

Quadrilaterals

A *convex quadrilateral ABCD* is the union of four segments (sides) AB, BC, CD, and DA that are placed in such a way that each side determines a half-plane that contains the opposite side (See Figure 7.18.) Note that $\overleftrightarrow{AC}$ intersects $\overleftrightarrow{BD}$ at an interior point of the figure. The other diagonal points (in the sense of projective geometry) can be distributed in six distinct configurations as far as incidence is concerned. This together with the possibilities for equality of various lengths and angles gives us a rich variety of generalizations of the notions of parallelogram, rectangle, rhombus, and square. We will only scratch the surface of this wealth of symmetric figures.

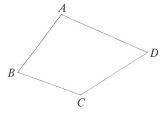

Figure 7.18   A convex quadrilateral.

First, consider a convex quadrilateral $ABCD$ in which opposite sides have the same length. This is the best generalization of the Euclidean notion of parallelogram. The special case in which all four sides have equal length is called a *rhombus*. (See Figure 7.19.)

A convex quadrilateral in which all four angles have equal radian measure is called an *equiangular quadrilateral*. The equiangular rhombus is the hyperbolic analogue of the square. (See Figures 7.20 and 7.21.)

Although convex quadrilaterals cannot have four right angles in hyperbolic geometry, there are several different figures that could be considered analogues of the rectangle. The *Saccheri quadrilateral ABCD* has $d(A, B) = d(C, D)$, $AB \perp BC$, and $CD \perp BC$. (See Figure 7.22.) The *Lambert quadrilateral*, on the other hand, has three right angles. It is shown in Figure 7.23.

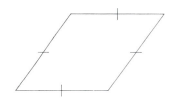

Figure 7.19   A rhombus.

175

Regular polygons

A regular polygon with any number of sides can be constructed by taking as vertices the orbit of a point Q under a cyclic subgroup of the group of rotations leaving another point P fixed. The resulting figure has the same symmetry group as in the Euclidean case. However, as the trigonometric formulas show, the angles get smaller as $d(P, Q)$ increases. In general, there are regular m-gons whose angles all have radian measure equal to any number between 0 and $(1 - 2/m)\pi$ that you wish to prescribe. For example, there is a regular 7-gon all of whose angles are right angles. See Figure 7.24.

Figure 7.20 An equiangular quadrilateral.

Congruence theorems

Theorem 44. *There is a unique reflection that interchanges a given pair of points of* $\mathbf{H}^2$.

Proof: Let P and Q be the given points. Let m be the perpendicular bisector of PQ. Note that the midpoint M of PQ is a unit timelike vector in the direction $[P + Q]$ and that the unit normal to m has direction $[P - Q]$. It is a straightforward exercise (Exercise 51) to verify that Ω_m interchanges P and Q. On the other hand, if $\Omega_{m'}$ is any reflection interchanging P and Q, then $\Omega_m\Omega_{m'}$ must leave P and Q fixed and, hence, by Theorem 36, must be the identity. □

Figure 7.21 An equiangular rhombus.

Theorem 45. *There are precisely two reflections that interchange a given pair of intersecting lines of* $\mathbf{H}^2$.

Proof: Suppose that the two lines have unit normals ξ and η. Let α be the line with unit normal in the direction $[\xi + \eta]$. It is easy (Exercise 52) to verify that Ω_α interchanges the two given lines. On the other hand, if $\Omega_{\alpha'}$ is any other reflection interchanging the given lines, the rotation $\Omega_\alpha\Omega_{\alpha'}$ leaves both lines fixed. By Theorem 37, α and α' must be perpendicular. Note that α' is just the line whose unit normals have direction $[\xi - \eta]$. □

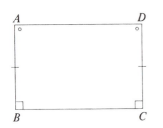

Figure 7.22 A Saccheri quadrilateral.

Classification of isometries of $\mathbf{H}^2$

Our main result is that every isometry of $\mathbf{H}^2$ is a motion. First, we have the following uniqueness theorems.

Theorem 46. *Let T be an isometry that leaves fixed a point P and a line ℓ through P. Let m be the line through P perpendicular to ℓ. Then either T or $\Omega_m T$ has ℓ as a line of fixed points.*

Proof: Let X be an arbitrary point of ℓ other than P. Let v be the unit direction vector of $\overrightarrow{PX}$. Then for some positive number s,

$$X = (\cosh s)P + (\sinh s)v.$$

Similarly, if Y is a third point on ℓ,

$$Y = (\cosh t)P + (\sinh t)v$$

for some t. Because $T\ell = \ell$, the points TX and TY have similar representations. Using the fact that $b(TX, P) = b(X, P)$ and $b(TY, P) = b(Y, P)$, we see that these representations take the form

$$TX = (\cosh s)P \pm (\sinh s)v, \quad TY = (\cosh t)P \pm (\sinh t)v. \quad (7.16)$$

Figure 7.23 A Lambert quadrilateral.

But now, $b(TX, TY) = b(X, Y)$, and, hence, the signs occurring in (7.16) are either both positive or both negative. In the first case T leaves ℓ pointwise fixed. It is easy to check that $\Omega_m T$ has the same property in the second case. $\qquad\square$

Theorem 47. *Let T be an isometry of **H**^{2}. Suppose that T has a line ℓ of fixed points. Then either $T = \Omega_\ell$ or T is the identity.*

Proof: Assume that T is not the identity. Choose any point X not fixed by T. Let A be the foot of the perpendicular from X to ℓ, and let v be the unit direction vector of $\overrightarrow{AX}$. We may then construct an orthonormal basis $\{e_1, e_2, e_3\}$ with $e_3 = A$, $e_2 = v$, and $e_1 = e_2 \times e_3$. Write

$$X = (\cosh t)e_3 + (\sinh t)e_2, \quad t > 0.$$

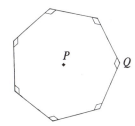

Figure 7.24 A regular 7-gon with seven right angles.

Choose $s > 0$ and consider on ℓ the points

$$Y = (\cosh s)e_3 + (\sinh s)e_1, \quad Y' = (\cosh s)e_3 - (\sinh s)e_1.$$

Using the fact that $d(X, Y) = d(X, Y')$ and, hence, $d(TX, Y) = d(TX, Y')$, we conclude that $b(TX, e_1) = 0$. Thus, TX must lie on the line $\overleftrightarrow{AX}$. Writing

$$TX = (\cosh u)e_3 + (\sinh u)e_2,$$

and using the fact that $b(TX, A) = b(X, A)$, we get $\cosh s = \cosh u$; that is, $s = \pm u$. From this it is clear that $TX = \Omega_\ell X$ and that T agrees with Ω_ℓ at every nonfixed point of T. It remains only to show that all fixed points of T lie on ℓ. To see this, suppose that Y is a fixed point of T. Then

$$\begin{aligned} b(X, Y) &= b(TX, TY) = b(\Omega_\ell X, Y) \\ &= b(X - 2b(X, e_2)e_2, Y) \\ &= b(X, Y) - 2b(X, e_2)b(Y, e_2). \end{aligned}$$

Because $b(X, e_2) \neq 0$, we must have $b(Y, e_2) = 0$; that is, Y lies on ℓ. $\qquad\square$

Theorem 48. *Every isometry of the hyperbolic plane is a motion.*

Proof: Let T be an isometry. We shall construct a sequence of reflections whose product coincides with T. First choose an arbitrary point P, and let T_1 be the reflection that interchanges P and TP. Then T_1T has P as a fixed point (Theorem 44). (In the special case where T already has a fixed point, we may shorten the construction by letting P be the fixed point and by letting $T_1 = I$.)

The second step is to construct T_2, so that T_2T_1T fixes P and a line ℓ through P. This may be arranged by letting T_2 be a reflection interchanging an arbitrary line ℓ through P with $T_1T\ell$ (Theorem 45). (If T_1T already has a fixed line ℓ through P, we may choose $T_2 = I$.)

Now, directly applying Theorem 46, we can choose a suitable reflection T_3 (or possibly $T_3 = I$), so that $T_3T_2T_1T$ leaves ℓ pointwise fixed. Finally, by Theorem 47, we can choose $T_4 = \Omega_\ell$ or $T_4 = I$, so that $T_4T_3T_2T_1T$ is the identity. Because each T_i is its own inverse, this means that $T = T_1T_2T_3T_4$, as required. $\qquad\square$

Remark: As we saw in Theorem 35, this product may be written as the product of three or fewer reflections. In this case, however, we can observe this fact more directly as follows. Using Theorem 17, we can see that $T_2T_3T_4$ is either a rotation about P or a reflection in a line through P. Then, depending on T_1, we can conclude that T is a rotation, a reflection, or a glide reflection. Furthermore, we have all the information necessary to explicitly find the transformations T_i.

Corollary. *Every isometry of* $\mathbf{H}^2$ *is one of the following: reflection, rotation, parallel displacement, translation, or glide reflection.*

Circles, horocycles, and equidistant curves

Definition. *Let C be a point and $r \geqslant 0$ a number. Then*

$$\mathscr{C} = \{X | d(X, C) = r\} \tag{7.17}$$

is called a circle *with center C and radius r.*

Theorem 49. *Let $\mathscr{P}$ be the pencil of lines through a point C, and let P be any point. Then the orbit of P by $\mathrm{REF}(\mathscr{P})$ is the circle with center C and radius $r = d(P, C)$. Conversely, every circle arises in this way.*

Definition. *Let m be a line and r a positive number. The portion of*

$$\{X | d(X, m) = r\} \tag{7.18}$$

lying in a half-plane determined by m is called an equidistant curve. *The line m is also (by definition) an equidistant curve corresponding to $r = 0$.*

Theorem 50. *Let $\mathscr{P}$ be the pencil of lines perpendicular to a line m, and let P be any point. Then the orbit of P by* REF$(\mathscr{P})$ *is an equidistant curve. Conversely, every equidistant curve arises in this way.*

Definition. *Let $\mathscr{P}$ be a pencil of parallels, and let P be any point. Then the orbit of P by* REF$(\mathscr{P})$ *is called a* horocycle.

Remark: The horocycle may be thought of as a limiting case of a circle having its center "at infinity."

Theorem 51. *Let v be a nonzero vector in $\mathbf{R}^3$, and let a be a number. If*

$$\{x \in \mathbf{H}^2 | b(v, x) = a\} \tag{7.19}$$

is nonempty, it is a circle, an equidistant curve, or a horocycle. Conversely, each circle, equidistant curve, and horocycle has an equation of this form.

A higher-dimensional version of the results of this section is found in [6].

EXERCISES

1. Prove Theorem 2.
2. Find spacelike vectors ξ and η such that $\xi \times \eta$ is a nonzero lightlike vector, but $b(\xi, \eta)^2 = b(\xi, \xi)b(\eta, \eta)$.
3. Verify that neither parallel nor ultraparallel lines intersect. (See the remark following Theorem 6.)
4. Verify the remarks following the definition of pencils.
5. i. Let $\xi = (1/\sqrt{2})(1, 1, 0)$ and $\eta = (1/2\sqrt{3})(3, 2, 1)$. Find an orthonormal basis with ξ as one element and a multiple of $\xi \times \eta$ as another.
   ii. Let ξ and η be the respective unit normals of lines of $\mathbf{H}^2$. If the lines intersect, find the point of intersection. If they are ultraparallel, find the common perpendicular.
6. i. Prove Theorem 9.
   ii. Prove that $d(\ell, m) = \cosh^{-1}|\langle \xi, \eta \rangle|$, where ℓ and m are lines with unit normals ξ and η.
7. Complete the proof of Theorem 11 by showing that $d(Q, R) = d(P, R) + d(P, Q)$ implies that P lies between Q and R.
8. Prove Theorem 12.
9. Verify that Theorems 17 and 18 hold.
10. Prove Theorem 19.
11. Prove Theorem 20 and the remark following it.

179

12. Prove Theorem 21.

13. Prove Theorem 22.

14. Prove Theorem 23 and its corollary.

15. Prove Theorem 24.

16. Check that $\pm\zeta$ must have the form $(1, r, -r)$ as indicated in Theorem 25.

17. Verify formula (7.7).

18. Prove Theorem 27.

19.   i. Check that $\mathrm{DIS}(\mathscr{P})$ is a subgroup of $\mathrm{REF}(\mathscr{P})$.
  ii. Show that if α is any line of the pencil $\mathscr{P}$, $\mathrm{REF}(\mathscr{P}) = \mathrm{DIS}(\mathscr{P}) \cup \{\Omega_\alpha D \,|\, D \in \mathrm{DIS}(\mathscr{P})\}$.
  iii. Check that the mapping $h \to D_h$ is an isomorphism of $\mathbf{R}$ (the additive group of real numbers) onto $\mathrm{DIS}(\mathscr{P})$.

20. Verify formula (7.9).

21. Prove Theorems 28 and 29.

22. Let ℓ be the common perpendicular of an ultraparallel pencil $\mathscr{P}$.
  i. Check that $\mathrm{TRANS}(\ell)$ is a subgroup of $\mathrm{REF}(\mathscr{P})$.
  ii. Show that if α is any line of the pencil $\mathscr{P}$,

$$\mathrm{REF}(\mathscr{P}) = \mathrm{TRANS}(\ell) \cup \{\Omega_\alpha \circ T \,|\, T \in \mathrm{TRANS}(\ell)\}.$$

    This means that $\mathrm{TRANS}(\ell)$ is a subgroup of index 2. One coset is $\mathrm{TRANS}(\ell)$, and the other is the set of reflections.
  iii. Check that the mapping $h \to T_h$ is an isomorphism of $\mathbf{R}$ onto $\mathrm{TRANS}(\ell)$.

23. If ℓ, m, and n are lines of a pencil, prove that $\Omega_\ell \Omega_m \Omega_n = \Omega_n \Omega_m \Omega_\ell$.

24. If H_1, H_2, and H_3 are distinct half-turns, prove that

$$H_1 H_2 H_3 \neq H_3 H_2 H_1.$$

25. If $T \in \mathrm{TRANS}(m)$ and $\ell \perp m$, show that $\Omega_\ell T = T\Omega_\ell$. Verify formula (7.10).

26. Using the matrix representation (7.10), show that a nontrivial glide reflection i. has no fixed points,
  ii. leaves fixed its axis and no other lines.

27. Prove that there is a unique reflection $\Omega_{p,q}$ interchanging any two lines p and q of a pencil of parallels (respectively, ultraparallels).

28. What is the square of a glide reflection in $\mathbf{H}^2$?

29. Describe the product of two glide reflections in $\mathbf{H}^2$ with perpendicular axes.

30. Let P and Q be points. Show that there is a unique translation taking P to Q.

31. Show that two nontrivial rotations of $\mathbf{H}^2$ commute if and only if they have the same center.

32. Let P, Q, and R be three noncollinear points of $\mathbf{H}^2$. Discuss the product of the half-turns H_P, H_Q, and H_R. Given a rotation, show that it can be expressed as the product of three half-turns.

33. Let γ be a line. Explain why no line can be both parallel to γ and perpendicular to γ.

34. Let v and w be nonproportional lightlike vectors. Prove that $b(v, w) \neq 0$.

35. Prove Theorem 34.

36. Let P, Q, and R be three points lying, respectively, on three members p, q, and r of a pencil of parallels. If P and Q are interchanged by Ω_{pq}, and Q and R are interchanged by Ω_{qr}, prove that
    i. P, Q, and R cannot be collinear.
    ii. Ω_{pr} interchanges P and R.
    (Notation is as in Exercise 27.)

37. Prove Theorem 36.

38. Fill in the missing details in the proof of Theorem 38.

39. Prove Theorem 39.

40. Prove that a segment AB consists of A, B and all points between A and B.

41. Verify the statements made in the text about the definition of radian measure of an angle.

42. Prove Theorem 40.

43. Prove the crossbar theorem in $\mathbf{H}^2$.

44. Prove the identity (7.12) for $a \in [-1, 1]$ and $\lambda, \mu \in (0, \infty)$.

45. Prove the formulas of hyperbolic trigonometry (Theorem 43).

46. Let ABC be a triangle in $\mathbf{H}^2$ with sides of lengths $a = d(B, C)$, $b = d(A, C)$, and $c = d(A, B)$. Prove that if AC is perpendicular to AB, then

$$\cosh a = \cosh b \cosh c.$$

   Find a direct proof that does not make use of Exercise 45.

47. The angle sum for a triangle in $\mathbf{H}^2$ is less than π. Prove this for the special cases of an equilateral triangle and a right-angled triangle.

48. The *defect* of a triangle in $\mathbf{H}^2$ is the amount by which its angle sum differs from π. Let $\triangle ABC$ be a triangle, and let F be a point between A and C that is the foot of the perpendicular from B to $\overleftrightarrow{AC}$. Prove that

$$\text{defect}(\triangle ABF) + \text{defect}(\triangle CBF) = \text{defect}(\triangle ABC).$$

Hence, prove that the defect of any triangle in $\mathbf{H}^2$ is positive.

Remark: The analogous conclusion can be drawn in spherical geometry or elliptic geometry. The angle sum is greater than π, and the amount of the difference is called the *excess*. Defect and excess can be used as measures of area in non-Euclidean geometry. The excess cannot be greater than 2π, and in fact there are spherical triangles whose areas are as close to 2π as we please. On the other hand, the defect of a triangle in $\mathbf{H}^2$ is less than π, and there are triangles whose areas are as close to π as we please.

49. Draw some pictures indicating how four points $ABCD$ might not determine a convex quadrilateral.

50. Show that a Saccheri quadrilateral can be decomposed into two Lambert quadrilaterals.

51. Fill in the missing details in the proof of Theorem 44.

52. Verify that the reflection Ω_α in Theorem 45 interchanges the two given lines.

53. i. Prove that there is a unique reflection interchanging any two distinct rays with common origin.
    ii. Prove that there are exactly two reflections interchanging two intersecting lines.

54. Let AB, BC, and CD be three line segments with $AB \perp BC$ and $BC \perp CD$. Given that AB and CD have equal length, prove that $d(A, C) = d(B, D)$. Work in $\mathbf{H}^2$, although your results should be equally valid in $\mathbf{E}^2$.

55. Find the symmetry group of
    i. the rhombus,
    ii. the equiangular quadrilateral,
    iii. the equiangular rhombus,
    iv. the Saccheri quadrilateral.

56. Formulate Hjelmslev's theorem so that it makes sense in $\mathbf{H}^2$. Is it true?

57. Verify that the SSS, SAS, and AAA congruence theorems hold in $\mathbf{H}^2$.

58. What congruence theorems hold for asymptotic triangles?

59. Verify that the concurrence theorems (4.53 and 4.54) are valid in the hyperbolic plane.

60. Let PQ and PX be perpendicular segments. Show that there is a unique ray $\overrightarrow{QY}$ such that PQ, $\overrightarrow{PX}$, and $\overrightarrow{QY}$ form an asymptotic triangle. If the radian measure of $\angle Q$ is θ and the length of PQ is d,

show that $\sin \theta \cosh d = 1$. The number θ is called the "angle of parallelism" determined by d. See Figure 7.25.

61. Prove Theorem 49.

62. Prove that a circle has only one center and one radius.

63. Discuss the various ways in which a circle can intersect a line or another circle.

64. Prove Theorem 50.

65. Prove that an equidistant curve uniquely determines the line m and the number r in (7.18).

66. Prove that a line meets an equidistant curve in at most two points.

67. Prove that a line meets a horocycle in at most two points.

68. Prove Theorem 51.

69. Identify the following curves in $\mathbf{H}^2$.
    i. $x_1 + x_2 = \sqrt{2} \sinh(2)$.
    ii. $x_3 = 2$.
    iii. $x_1 + x_3 = 2$.

70. Prove that the groups $\mathrm{ROT}(\mathscr{P})$, $\mathrm{TRANS}(m)$, and $\mathrm{DIS}(\mathscr{P})$ would have worked equally well in characterizing circles, equidistant curves, and horocycles, respectively. What are the stabilizers in these cases?

71. Investigate the status of Theorems 49–50 and Exercises 61–66 in the Euclidean, spherical, and projective settings.

72. Investigate the relationships suggested in the remark following Theorem 43.

Circles, horocycles, and equidistant curves

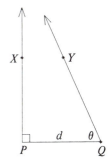

Figure 7.25 The angle of parallelism.

183

APPENDIX A The axiomatic approach

In this book we have approached plane geometry from the analytic point of view. The more traditional treatment of these topics (going back to Euclid) is based on geometric axioms or postulates and synthetic proofs (i.e., not using computation).

This approach is also instructive and complements our own. Essentially the same set of theorems can be derived, and the reader is encouraged to take the exercises from a book using the axiomatic approach and try to solve them by using our methods.

An excellent book using the axiomatic approach is that of Greenberg [16]. Following Hilbert, whose work we mentioned in the Historical Introduction, he first introduces incidence axioms, which guarantee that two points are incident with a unique line, that every line is incident with at least two points, and that not all points are collinear. Then he introduces axioms for betweenness from which segment, ray, half-plane, angle, and its interior can be defined. These axioms are strong enough to allow one to prove Pasch's theorem and the crossbar theorem. The betweenness axioms are already too strong to be consistent with the geometry of the sphere or the projective plane, however. (A different set of betweenness axioms would be necessary.) Then an (undefined) relation of congruence on segments, angles, and triangles is introduced along with certain axioms that congruence is to satisfy. Finally, continuity axioms are stated that essentially guarantee that lines behave like the "real number line" of analytic geometry. This is enough to force the geometry being discussed to be either the Euclidean or hyperbolic plane, depending on whether or not Euclid's fifth postulate is satisfied. This approach is ideal for investigating the dependence of geometrical results on the parallel postulate. Absolute geometry can be derived first; then, by appropriate choice of parallel axiom, Euclidean or hyperbolic geometry may be derived. In this development, however, it is not easy to get to the point of dealing with isometries or to compute with them once they have been defined.

A second approach, also axiomatic, has been taken by Ewald [12]. He also begins with incidence axioms but then introduces the (undefined) symmetric relation of perpendicularity in the set of lines. Perpendicularity axioms equivalent to our Theorems 1.13 and 1.14 are given, except that the result of dropping a perpendicular is not assumed to be unique. Next, the notion of reflection is introduced. It is a collineation that preserves perpendicularity and has a line of fixed points. Each line is assumed to be the axis of exactly one reflection. One can consider two types of pencils – the lines through a certain point and the lines perpendicular to a certain line – and, hence, define rotation and translation. Several additional reflection axioms are stated, essentially guaranteeing that two lines that are related by a rotation or translation are also related by a reflection. This axiom system is general enough to allow the Euclidean, elliptic, and hyperbolic planes as models.

The three reflections theorem and representation theorem are satisfied for both types of pencils. If the two types of pencils coincide, the geometry is elliptic. If not, it may be hyperbolic or Euclidean, depending on whether or not there exist two nonintersecting lines without a common perpendicular. In this latter case a third type of pencil can be defined merely by requiring that the three reflections theorem be satisfied. All the usual properties of pencils of parallels in hyperbolic geometry follow. Adding suitable continuity axioms now forces the geometry to be (up to a normalization factor) the Euclidean, elliptic, or hyperbolic plane studied in the text.

Sets and functions

If A is a set, the number of elements (cardinality) of A is denoted by $\#A$. This number is allowed to be a nonnegative integer or ∞. We will not be concerned with different cardinalities of infinite sets.

Let A and B be sets. A subset f of $A \times B$ is called a *function* from A to B if every element of A occurs exactly once as a first member of a pair in f. We use the notation

$$f: A \rightarrow B$$

to indicate that f is a function from A to B. The set A is called the *domain* of f. (Functions are sometimes called *mappings*.)

For any element x of A there is a unique element y of B such that $(x, y) \in f$. Thus, the function may also be thought of as a means by which each element x in A determines a unique element of B. This element is usually denoted by $f(x)$. We say that f maps x to $f(x)$ or that $f(x)$ is the *image* of x under f. This terminology extends to sets as well. If S is a subset of A, the image of S under f is denoted by $f(S)$ and is defined to be

$$f(S) = \{f(x) \mid x \in S\}.$$

The function f is said to be *surjective* if $f(A) = B$; that is, every element of B occurs in the image of A. In this case we also say that f maps A *onto* B. The function f is said to be *injective* if no two elements of A map to the same element of B. A function that is both surjective and injective is said to be a *bijective* function or a *bijection*. A bijection provides a matching or one-to-one correspondence between the sets A and B.

Notation: We sometimes omit the parentheses and write fx for $f(x)$ and fS for $f(S)$.

Composition and inverse functions

Suppose we have $f: A \rightarrow B$ and $g: B \rightarrow C$. Then the *composite function*

$$g \circ f: A \rightarrow C$$

consists of those pairs (x, z) in $A \times C$ such that $z = g(f(x))$.
If f is a bijection, then

$$g = \{(y, x)|(x, y) \in f\}$$

is also a bijection. It is easy to check that both $g \circ f$ and $f \circ g$ are identity functions on their respective domains. In this case we say that f and g are inverses of each other and write

$$f^{-1} = g \quad \text{and} \quad g^{-1} = f.$$

Identity functions will be denoted by I. The domain should be clear from the context.

Permutations

A bijection from a set A to itself is called a *permutation* of A. Such a bijection can be composed with itself any number of times. This leads to the following algebraic notation:

$$f^1 = f, \quad f^2 = f \circ f, \quad f^n = f \circ f^{n-1}. \tag{B.1}$$

This notation may be extended to include zero and negative exponents by setting

$$f^0 = I$$

and

$$f^{-n} = (f^{-1})^n. \tag{B.2}$$

A function satisfying $f^2 = I$ (but $f \neq I$) is called an *involution*.

Permutations are sometimes expressed in cyclic notation. Let $\{x_i\}$, $1 \leqslant i \leqslant n$, be a set of objects. Then the permutation that maps x_n to x_1 and all other x_i to x_{i+1} is denoted by $(x_1 x_2 \cdots x_n)$. Such a permutation is called a *cycle* or *cyclic* permutation. This notation is used in Chapter 2 when discussing symmetries of a triangle. Note that (PQR), (QRP), and (RPQ) all denote the same permutation, and (PRQ), (QPR), and (RQP) denote its inverse.

A cycle of length 2, such as (PQ), is called a *transposition*. Every permutation of a finite set can be expressed as a product of transpositions. If s is the number of transpositions, then $(-1)^s$ is called the *sign* of the permutation. For example, the sign of (PQ) is -1, and the sign of $(PQR) = (PQ)(QR)$ is $+1$.

Relations

Let A be a set. A relation on the set A is a subset of $A \times A$. A relation r on A is said to be

i.   *reflexive* if it contains all possible pairs of the form (a, a);

ii. *symmetric* if for each element (x, y) of r, the reversed pair (y, x) is also in r;

iii. *transitive* if whenever (x, y) and (y, z) are elements of r, (x, z) is also a member of r.

An *equivalence relation* is a relation that is reflexive, symmetric, and transitive. Let r be an equivalence relation. For each element a in A, let

$$[a] = \{x \in A | (x, a) \in r\}.$$

Then one can verify that any two distinct sets of the form $[a]$ must be disjoint and that the union of all such sets is A. Thus, an equivalence relation partitions A into subsets called *equivalence classes*. The set of equivalence classes is called the *quotient space A/r*. The function that sends each element of A to the equivalence class to which it belongs is called the *natural projection*. The notion of equivalence relation is used several places in the text – in particular, when discussing groups and the projective plane.

Groups

A *group* is a set G together with a function from $G \times G$ to G satisfying certain conditions. In order to express these conditions in a compact way, we adopt the following notation. If a and b are elements of G, we denote by $a * b$ the result of applying the function to the ordered pair (a, b). The conditions for G to be a group are

1. $(a * b) * c = a * (b * c)$ for all a, b, and c in G. This is called the *associative law*.
2. There is an element I of G such that $a * I = I * a = a$ for all a in G. Such an element is called an *identity*.
3. For each element a in G there is an element b in G such that $a * b = b * a = I$. The elements b and a are said to be *inverses* of each other.

Remark: It is not difficult to show that there is only one identity element in a group and that each element has exactly one inverse.

When discussing group operations, we often use informal terminology that exploits the analogy with mutiplication of numbers. For instance, $a * b$ is sometimes called the *product* of a and b. This should not mislead us into thinking that $*$ possesses all the properties of ordinary multiplication. For instance, the *commutative law* $a * b = b * a$ is usually false in the groups we will be using. If two elements do satisfy this condition, we say that they *commute* with each other. If this commutative law is satisfied for all pairs of elements in the group, the group itself is said to be commutative or *abelian*.

It is common when dealing with groups to omit explicit mention of the operator $*$ when the meaning is clear from the context. For example, the commutative law might be written $ab = ba$. Also, the constructions of (B.1) and (B.2) may be applied to any element of any group (e.g., $a^2 = aa$, $a^0 = I$, a^{-1} is the inverse of a, etc.).

Although the notion of group is a very general one and admits many interpretations, the original motivation for the definition and the interpretation with which we will be primarily concerned is that of *groups of transformations*. By this we mean that group elements are bijections of some set to itself, and the group operation is composition of functions.

Theorem 1C. *The bijections of any set to itself form a group.*

Subgroups

Let G be a group as previously defined. A subset H of G is said to be a *subgroup* of G if the following conditions are satisfied:

1. If $a, b \in H$, then $a * b \in H$ (the *closure* property).
2. The identity I of G is in H.
3. If $a \in H$, then $a^{-1} \in H$.

Informally, this means that the subset H with the operation inherited from G is itself a group. Of course, property (2) follows from (1) and (3).

Remark: The intersection of any collection of subgroups of a group is a subgroup.

Let G be a group and S any subset of G. The intersection of all subgroups containing S is called the *subgroup generated by S* and is denoted by $\langle S \rangle$. If S consists of a single element s, we use the notation $\langle s \rangle$.

Theorem 2C. *Every element of $\langle S \rangle$ can be written as the product of a finite number of elements, each of which is either an element of S or the inverse of an element of S.*

An element s of a group G is said to be of order n if $\#\langle s \rangle = n$. A group G is of order n if $\#G = n$.

Cosets

If H is a subgroup of a group G, and a is any element of G, we define the *left coset aH* to be

$$aH = \{ah | h \in H\}.$$

Theorem 3C.
i. *Any two distinct left cosets are disjoint.*
ii. *Two left cosets aH and bH coincide if and only if $a^{-1}b \in H$.*

Remark: $\{(a, b) | aH = bH\}$ is an equivalence relation, and the cosets are the equivalence classes. (See Appendix B.)

Because G is the union of its left cosets and all left cosets have the same cardinality, we have the following.

Corollary. *Let G be a finite group, and let H be any subgroup. Then #H divides #G. (Their quotient is the number of cosets and is called the index of H in G and denoted by $[G:H]$.)*

Remark: Right cosets are defined analogously, and similar conclusions hold.

Homomorphisms and isomorphisms

Let G_1 and G_2 be groups. A function

$$f: G_1 \to G_2$$

is called a *homomorphism* if for all $a, b \in G_1$,

$$f(a * b) = f(a) * f(b);$$

that is, f respects the group structure. From this property it is not hard to show that a homomorphism also satisfies $f(I) = I$ and $f(a^{-1}) = (f(a))^{-1}$ for all $a \in G_1$.

A bijective homomorphism is called an *isomorphism*, and the groups involved are said to be *isomorphic*.

Note that G_1 and G_2 are different groups, each having its own group operation and identity element even though we have used the same symbol.

Definition. *Let f be a homomorphism of groups. Then*

$$\{a \mid f(a) = I\}$$

is called the kernel *of f (abbreviated ker f).*

Theorem 4C. *Let $f: G_1 \to G_2$ be a homomorphism. Then ker f is a subgroup of G_1, and $f(G_1)$ is a subgroup of G_2.*

Quotient groups and normal subgroups

Definition. *A subgroup H of a group G is said to be a* normal *subgroup if $aH = Ha$ for all a in G.*

If H is a normal subgroup, the cosets form a group with the operation

$$(aH) * (bH) = (a * b)H.$$

In checking that this operation on cosets is well-defined, we must use the fact that H is normal.

The group of cosets is called the *quotient group* of G by H and is denoted by G/H. The function π that takes each $a \in G$ to the coset aH to which it belongs is a homomorphism called the *natural projection* of G onto the quotient group G/H.

Theorem 5C. *Let $f: G_1 \to G_2$ be a surjective homomorphism. Let H be the kernel of f. Then the mapping given by*

$$\tau(aH) = f(a)$$

is an isomorphism of G_1/H to G_2. Furthermore, $\tau \circ \pi = f$.

Notation: The statement "G_1 is isomorphic to G_2" is abbreviated $G_1 \cong G_2$.

Linear algebra

Linear algebra for $\mathbf{R}^2$

Let A be a 2 × 2 matrix. Define T_A: $\mathbf{R}^2 \rightarrow \mathbf{R}^2$ by

$$T_A x = Ax, \tag{D.1}$$

where the right side is matrix multiplication of the 2 × 2 matrix A by the 2 × 1 matrix (column vector) x. One can verify that the following properties hold for all $x, y \in \mathbf{R}^2$ and all $r \in \mathbf{R}$,

1. $T_A(x + y) = T_A x + T_A y$, $\qquad$ (D.2)
2. $T_A(rx) = rT_A x$.

In other words, T_A is *linear*. Conversely, let f: $\mathbf{R}^2 \rightarrow \mathbf{R}^2$ be linear. Each element $x \in \mathbf{R}^2$ may be written $x = x_1 \varepsilon_1 + x_2 \varepsilon_2$, so that

$$fx = f(x_1\varepsilon_1) + f(x_2\varepsilon_2) = x_1 f\varepsilon_1 + x_2 f\varepsilon_2.$$

Let

$$f\varepsilon_1 = \begin{bmatrix} a_{11} \\ a_{21} \end{bmatrix} \quad \text{and} \quad f\varepsilon_2 = \begin{bmatrix} a_{12} \\ a_{22} \end{bmatrix}.$$

Then

$$fx = \begin{bmatrix} x_1 a_{11} \\ x_1 a_{21} \end{bmatrix} + \begin{bmatrix} x_2 a_{12} \\ x_2 a_{22} \end{bmatrix} = \begin{bmatrix} a_{11}x_1 + a_{12}x_2 \\ a_{21}x_1 + a_{22}x_2 \end{bmatrix}$$

$$= \begin{bmatrix} a_{11} & a_{12} \\ a_{21} & a_{22} \end{bmatrix}\begin{bmatrix} x_1 \\ x_2 \end{bmatrix} = T_A x,$$

where

$$A = \begin{bmatrix} a_{11} & a_{12} \\ a_{21} & a_{22} \end{bmatrix}.$$

Thus, for each linear function f: $\mathbf{R}^2 \rightarrow \mathbf{R}^2$ there is a matrix A such that $f = T_A$. Furthermore, A is uniquely determined by T_A. To see this, note

$$T_A \varepsilon_1 = \begin{bmatrix} a_{11} & a_{12} \\ a_{21} & a_{22} \end{bmatrix} \begin{bmatrix} 1 \\ 0 \end{bmatrix} = \begin{bmatrix} a_{11} \\ a_{21} \end{bmatrix},$$

$$T_A \varepsilon_2 = \begin{bmatrix} a_{11} & a_{12} \\ a_{21} & a_{22} \end{bmatrix} \begin{bmatrix} 0 \\ 1 \end{bmatrix} = \begin{bmatrix} a_{12} \\ a_{22} \end{bmatrix}.$$

So, if $T_A = T_B$, we have

$$\begin{bmatrix} a_{11} \\ a_{21} \end{bmatrix} = \begin{bmatrix} b_{11} \\ b_{12} \end{bmatrix} \quad \text{and} \quad \begin{bmatrix} a_{12} \\ a_{22} \end{bmatrix} = \begin{bmatrix} b_{12} \\ b_{22} \end{bmatrix};$$

that is, $A = B$. Finally,

$$T_A T_B x = T_A B x = A(Bx) = (AB)x = T_{AB}x. \tag{D.3}$$

We summarize the above discussion as follows:

Theorem 1D.

i. *The mapping $A \to T_A$ that associates to each 2×2 matrix A a linear function $T_A \colon \mathbf{R}^2 \to \mathbf{R}^2$ is a bijection from the set of all such matrices to the set of all such linear functions. The columns of A are the images of the standard basis vectors ε_i under T_A.*

ii. *Composition of functions corresponds to matrix multiplication; that is, $T_A T_B = T_{AB}$.*

Corollary. *Matrix multiplication is associative.*

Proof: Consider the identity

$$(T_A T_B) T_C = T_A (T_B T_C),$$

which is true by definition of composition. This implies by (ii) of the theorem that

$$T_{AB} T_C = T_A T_{BC}, \quad T_{(AB)C} = T_{A(BC)}.$$

Because the matrix of a transformation is uniquely determined, we have

$$(AB)C = A(BC). \tag{D.4} \quad \square$$

Definition. *The set of linear functions from $\mathbf{R}^2 \to \mathbf{R}^2$ that are bijections is denoted by* $\mathbf{GL(2)}$.

Theorem 2D. $\mathbf{GL(2)}$ *is a subgroup of the group of all bijections of* $\mathbf{R}^2 \to \mathbf{R}^2$.

Proof: We need to show that

i. The composition of linear functions is linear,

ii. The inverse of a linear bijection is linear.

Because of the identity $T_A T_B = T_{AB}$, the first point is clear. Also, $T_I = I$, where the subscript I stands for the identity matrix.

Finding the inverse of a linear function is, in general, not an easy task. However, for a 2×2 matrix, there is no difficulty. In fact, we have the identity

$$\begin{bmatrix} a & b \\ c & d \end{bmatrix}\begin{bmatrix} d & -b \\ -c & a \end{bmatrix} = \begin{bmatrix} ad - bc & 0 \\ 0 & ad - bc \end{bmatrix}$$

and   (D.5)

$$\begin{bmatrix} d & -b \\ -c & a \end{bmatrix}\begin{bmatrix} a & b \\ c & d \end{bmatrix} = \begin{bmatrix} ad - bc & 0 \\ 0 & ad - bc \end{bmatrix}.$$

The quantity $ad - bc$ is called the *determinant* of the matrix. If the determinant is not zero, the inverse of the corresponding linear function is easily constructed (see the next corollary). If the determinant is zero, the function sends the vectors

$$\begin{bmatrix} d \\ -c \end{bmatrix} \quad \text{and} \quad \begin{bmatrix} -b \\ a \end{bmatrix}$$

to zero and thus cannot be injective.   □

Notation: The determinant of a matrix A is denoted by $\det A$.

Corollary. *Let A be a 2×2 matrix. Then*
i.   *T_A is a bijection if and only if $\det A \neq 0$.*
ii. *The matrix*

$$B = \begin{bmatrix} \dfrac{d}{\det A} & \dfrac{-b}{\det A} \\[2ex] \dfrac{-c}{\det A} & \dfrac{a}{\det A} \end{bmatrix}$$ (D.6)

satisfies

$$AB = BA = I,$$ (D.7)

and, hence, T_B is the inverse of T_A.
iii. **GL(2)** *is a group.*

Definition. *A set $\{v, w\}$ of vectors in $\mathbf{R}^2$ is a* basis *if for every vector $x \in \mathbf{R}^2$ there exist unique numbers λ, μ such that*

$$x = \lambda v + \mu w;$$ (D.8)

that is, x can be expressed uniquely as a linear combination *of v and w.*

To verify the uniqueness part of the definition in a particular case, it is only necessary to check the case $x = 0$. In fact, if

$$x = \lambda_1 v + \mu_1 w \quad \text{and} \quad x = \lambda_2 v + \mu_2 w,$$

we have

$$0 = (\lambda_1 - \lambda_2)v + (\mu_1 - \mu_2)w.$$

If we know that 0 can be represented in only one way, then we must have $\lambda_1 = \lambda_2$ and $\mu_1 = \mu_2$, so that x is also uniquely represented. We use this fact in the next theorem.

The complex structure J

The linear function $J: \mathbf{R}^2 \to \mathbf{R}^2$, defined by

$$J\begin{bmatrix} x_1 \\ x_2 \end{bmatrix} = \begin{bmatrix} -x_2 \\ x_1 \end{bmatrix},$$

is sufficiently important to deserve a special notation. It is used in Chapter 3.

J has the special property that $J^2 = -I$ and can be used to identify the matrix algebra of $\mathbf{R}^2$ with the algebra of the complex numbers. The notion also generalizes to higher dimensions, and any linear function satisfying $J^2 = -I$ is called a *complex structure*.

In $\mathbf{R}^2$, J coincides with rotation by $\pi/2$; that is, $J = \text{rot}(\pi/2)$ in the notation of Chapter 1. Also, $Jv = v^\perp$ for all vectors v, and, hence, J satisfies the identity $\langle Jv, v \rangle = 0$. In fact,

$$\langle Jv, w \rangle = -\langle v, Jw \rangle$$

for all vectors v and w.

Theorem 3D. *Let $\{v, w\}$ be a set of two vectors in $\mathbf{R}^2$. Let*

$$A = \begin{bmatrix} v_1 & w_1 \\ v_2 & w_2 \end{bmatrix}.$$

Then $\{v, w\}$ is a basis if and only if $\det A \neq 0$.

Proof: Suppose that $\det A \neq 0$. Equations (D.5) imply that

$$\lambda_1 v + \mu_1 w = \varepsilon_1, \quad \lambda_2 v + \mu_2 w = \varepsilon_2, \tag{D.9}$$

where

$$B = \begin{bmatrix} \lambda_1 & \lambda_2 \\ \mu_1 & \mu_2 \end{bmatrix}$$

is the inverse of A. Thus each of the standard basis vectors ε_1, ε_2 (and hence every vector) is a linear combination of $\{v, w\}$. Furthermore, any equation of the form

$$\lambda v + \mu w = 0$$

can be written as

$$A \begin{bmatrix} \lambda \\ \mu \end{bmatrix} = 0,$$

which gives

$$BA \begin{bmatrix} \lambda \\ \mu \end{bmatrix} = 0;$$

that is,

$$\begin{bmatrix} \lambda \\ \mu \end{bmatrix} = 0,$$

and $\{v, w\}$ is a basis.

Conversely, if $\det A = 0$, (D.5) shows that

$$w_2 v - v_2 w = 0 \quad \text{and} \quad -w_1 v + v_1 w = 0. \qquad \text{(D.10)}$$

This shows that there is more than one way of representing the zero vector as a linear combination of v and w, and, hence, that $\{v, w\}$ is not a basis. □

Definition. *Two vectors v and w are* proportional *if $v = \lambda w$ or $w = \mu v$ for some numbers λ and μ.*

Corollary. *A set $\{v, w\}$ of vectors in $\mathbf{R}^2$ is a basis if and only if v and w are not proportional.*

Proof: First, suppose that $v = kw$ for some number k. Then it is easy to check that $\det A = 0$ (where A is defined as in Theorem 3D). Conversely, if $\det A = 0$, (D.10) shows that v and w must be proportional. □

Linear algebra for $\mathbf{R}^n$

It would not be appropriate in this book to develop the fundamentals of linear algebra. The preceding section on the two-dimensional case can be approached from a naive point of view. It serves as a motivation and concrete realization of some of the more abstract concepts we need.

The reader of this section should already be familiar with the notions of vector space, linear function, linear dependence, and basis. Any linear algebra book covers these notions. The books of Hoffman–Kunze [19], Nomizu [26], and Banchoff–Wermer [3] are recommended.

For convenience we collect some definitions and facts here.

Definition. *Let S be a subset of a vector space V.*
i. *An element of V that is a sum of finitely many elements that are multiples of elements of S is said to be a* linear combination *from S. The set of such linear combinations is called the* span *of S and is denoted by* [S]. *A typical linear combination would have the form*

$$x = \sum_{i=1}^{m} \lambda_i v_i, \qquad (D.11)$$

where $\{v_i\} \subset S$ *and* $\{\lambda_i\} \subset \mathbf{R}$. *If the* λ_i *are not all zero, we speak of a* nontrivial *linear combination.*
ii. *S is a* linearly dependent *set if the zero vector can be written as a nontrivial linear combination from S. Otherwise, S is* linearly independent.

Remark:
i. If S contains the zero vector, then S is linearly dependent because $0 = \lambda_1 v_1$, where $\lambda_1 = 1$, $v_1 = 0$.
ii. If S contains two proportional vectors, then S is linearly dependent. If $v = kw$, then $v + (-k)w = 0$.

Definition. *A set S is a* basis *for V if every element of V can be expressed uniquely as a linear combination from S; that is, S is a linearly independent set spanning V.*

Definition. *Let n be a positive integer. A vector space V is said to be* n-dimensional *if it has a basis with n elements.*

Remark: For each positive integer n, $\mathbf{R}^n$ is an n-dimensional vector space. A basis is given by $\varepsilon_1 = (1, 0, 0, \ldots, 0)$, $\varepsilon_2 = (0, 1, 0, \ldots, 0)$, and so on. This is called the *standard basis of* $\mathbf{R}^n$.

Theorem 4D. *Let V be an n-dimensional vector space. Then*
i. *Every basis for V consists of n elements.*
ii. *Every linearly independent set of cardinality n is a basis.*
iii. *Every set of cardinality n that spans V is a basis.*
iv. *Every linearly independent set is a subset of some basis.*
v. *Every spanning set contains some basis.*

Matrices and linear functions

Let $f: V \to V$ be a linear function from an n-dimensional vector space V to itself. If $\{e_i\}$ is a basis of V, the matrix of f with respect to V is the matrix whose columns are the vectors $f(e_i)$ expressed with respect to the basis $\{e_i\}$. In other words, if

$$f(e_i) = \sum_{j=1}^{n} a_{ji}e_j, \qquad (D.12)$$

then the matrix $A = [a_{ij}]$ is the matrix of f. If $\{\tilde{e}_i\}$ is another basis for V, we may write

$$\tilde{e}_j = \sum_{i=1}^{n} p_{ij}e_i. \qquad (D.13)$$

If $\tilde{A}$ is the matrix of f with respect to $\{\tilde{e}_j\}$, then

$$\tilde{A} = P^{-1}AP. \qquad (D.14)$$

Remark: Let A be an $n \times n$ matrix. As before, set $T_A x = Ax$ for $x \in \mathbf{R}^n$. Then A is the matrix of T_A with respect to the standard basis $\{\varepsilon_i\}$. Theorems 1D and 2D hold true in the n-dimensional case as well.

The set of all linear bijective functions of $\mathbf{R}^n \to \mathbf{R}^n$ is a group denoted by **GL(n)** (the general linear group). The group of all invertible $n \times n$ matrices with matrix multiplication as the group operation is essentially the same thing. It is not so easy to write down the inverse of a matrix explicitly, however.

Multilinear functions

Definition. *Let V be a vector space, and let k be a positive integer. A k-linear function on V is a function*

$$f: V^k \to \mathbf{R}$$

that is linear in each variable separately. (V^k denotes the Cartesian product of k copies of V.) In other words, for each choice $(v_1, v_2, \ldots, v_k) \in V^k$ the mapping

$$v \to f(v_1, v_2, \ldots, v_{i-1}, v, v_{i+1}, \ldots, v_n)$$

is linear.

A function that is k-linear is sometimes said to be *multilinear of degree k.*

Linear algebra

A k-linear function is said to be *symmetric* if its values are unchanged when its arguments are permuted. It is said to be *alternating* if its values are unchanged in magnitude when its arguments are permuted but the signs change according to the sign of the permutation.

Remark: A multilinear function is determined by its values on any basis. For example, if f is a two-linear function on $\mathbf{R}^3$ and $\{e_1, e_2, e_3\}$ is a basis, the values of f are determined by the 3×3 matrix $[f(e_i, e_j)]$. If f is symmetric, then $f(e_1, e_2) = f(e_2, e_1)$, and so on, so the matrix is symmetric. On the other hand, if f is alternating, $f(e_1, e_1) = 0$, $f(e_1, e_2) = -f(e_2, e_1)$, and so on. The matrix is skew-symmetric.

Remark:
i. The inner products defined on $\mathbf{R}^2$ and $\mathbf{R}^3$ are examples of two-linear (or *bilinear*) functions. These functions are symmetric. Symmetric bilinear functions are also used to define polarities in Chapter 5 and the hyperbolic plane in Chapter 7.
ii. The determinant, considered as a function on the columns of a 2×2 matrix is bilinear and alternating. Because a bilinear, alternating function on $\mathbf{R}^2$ is determined by $f(\varepsilon_1, \varepsilon_2)$, the determinant can be characterized as the unique bilinear alternating function on $\mathbf{R}^2$ satisfying $f(\varepsilon_1, \varepsilon_2) = 1$.

Determinants

The preceding remark leads to the following definition.

Definition. *Let A be an $n \times n$ matrix. Let $\{v_1, v_2, \ldots, v_n\}$ be the columns of A. Then its determinant det A is the value on $(v_1, \ldots, v_n)$ of the unique n-linear alternating function that assigns the value 1 to $(\varepsilon_1, \varepsilon_2, \ldots, \varepsilon_n)$.*

Theorem 5D. *The determinant function satisfies the following identities:*
i. $\det(AB) = \det A \det B$.
ii. $\det(I) = 1$.
iii. $\det(A^{-1}) = 1/\det A$.
iv. $\det(A^t) = \det A$.

Proof: See Nomizu [26], Theorems 6.8 and 6.10. ☐

Corollary. *The determinant of the matrix of a linear function $T: V \to V$ is independent of the choice of basis. Thus, we can define* det *T without ambiguity.*

Theorem 6D.

i. *Let $T: V \to V$ be a linear function from an n-dimensional space V to itself. Then T is a bijection if and only if $\det T \neq 0$.*

ii. *Let A be an $n \times n$ matrix. Then the columns of A are a basis for $\mathbf{R}^n$ if and only if $\det A \neq 0$.*

Remark: This shows that $\mathbf{GL}(n)$ is the set of all linear mappings of $\mathbf{R}^n \to \mathbf{R}^n$ having nonzero determinant. The subset having determinant equal to 1 is a subgroup called the special linear group and denoted by $\mathbf{SL}(n)$. The group $\mathbf{SL}(3)$ is used in discussing the collineations of the projective plane in Chapter 5.

Eigenvectors and eigenvalues

In geometry we are often interested in computing fixed points and fixed lines of geometric transformations. In what follows, V is an n-dimensional vector space.

Definition. *Let $T: V \to V$ be linear. A nonzero vector $x \in V$ is said to be an eigenvector of T if Tx is proportional to x. The number λ such that $Tx = \lambda x$ is the associated eigenvalue.*

Note that $Tx = \lambda x$ is the same as $(T - \lambda I)x = 0$, where I is the identity function. We are thus led to consider the *characteristic polynomial* of T, namely,

$$\text{char}(t) = \det(T - tI). \tag{D.15}$$

One can verify that this is a polynomial of degree n in the variable t. To find an eigenvector x, find a value of t for which $\det(T - tI) = 0$. Then, because $T - tI$ is not injective, such an x exists.

In general, the theory of eigenvectors is complicated. Each root of the characteristic polynomial gives an eigenvalue. However, one eigenvalue can correspond to a many-dimensional space of eigenvectors. In our case, however (dimensions 2 and 3), we can explicitly determine the eigenvalues and eigenvectors without much trouble. (See Theorem 4.24.)

Representing linear functions

Let b be a bilinear function on $\mathbf{R}^n$. Choose any vector $\xi \in \mathbf{R}^n$ and consider the function

$$h_\xi: \mathbf{R}^n \to \mathbf{R}$$

defined by

$$h_\xi(x) = b(\xi, x). \tag{D.16}$$

Clearly, h_ξ is linear. In fact, this is (in part) what it means for b to be bilinear. Of course, h_ξ could be the zero function. This will happen obviously if $\xi = 0$.

Definition.   *The bilinear function b is* degenerate *if there is some nonzero ξ such that h_ξ is identically zero. Otherwise, b is said to be* nondegenerate.

Theorem 7D.   *Let $\{e_i\}$ be any basis for $\mathbf{R}^n$. If b is a nondegenerate bilinear function, then the matrix $B = [b(e_i, e_j)]$ is invertible.*

Proof: Suppose not. Then there are numbers $\xi_1, \ldots, \xi_n$ (not all zero) such that

$$\sum_{i=1}^{n} b(e_i, e_j)\xi_i = 0 \quad \text{for all } j;$$

that is, $b(\xi, e_j) = 0$ for all j. This implies that $h_\xi = 0$, contradicting nondegeneracy. $\square$

Theorem 8D.   *Let b be a nondegenerate bilinear function on $\mathbf{R}^n$. Let $f: \mathbf{R}^n \to \mathbf{R}$ be linear. Then there is a unique vector $\xi \in \mathbf{R}^n$ such that $f = h_\xi$; that is, $f(x) = b(\xi, x)$ for all $x \in \mathbf{R}^n$.*

Proof: Let $\{e_i\}$ be a basis for $\mathbf{R}^n$. Consider the equations

$$\sum b(e_j, e_i)\xi_j = f(e_i), \quad i = 1, 2, \ldots, n. \tag{D.17}$$

These equations may be rewritten $B\xi = \eta$, where $\eta = \sum_{i=1}^{n} f(e_i)\varepsilon_i$. But now using the fact that B is invertible, we have $\xi = B^{-1}\eta$. Thus, we have taken the original b and f, produced B and η, and $\xi = B^{-1}\eta$. It is easy to verify that with this value of ξ, $h_\xi = f$ as required.

For uniqueness, note that if $h_\xi = h_{\xi'}$, then $h_{\xi-\xi'} = 0$. By nondegeneracy, $\xi = \xi'$. $\square$

The results of this theorem are used in Theorem 4.1.

Proof of Theorem 2.2 APPENDIX E

In this appendix we provide a detailed proof that every (affine) collineation of $\mathbf{E}^2$ is an affine transformation. This is Theorem 2 of Chapter 2. In order to make this appendix self-contained, we restate the definitions.

Definition. *A* collineation *is a bijection* $T: \mathbf{E}^2 \rightarrow \mathbf{E}^2$ *satisfying the condition that for all triples* P, Q, *and* R *of distinct points,* P, Q, *and* R *are collinear if and only if* TP, TQ, *and* TR *are collinear.*

Definition. *A mapping* $T: \mathbf{E}^2 \rightarrow \mathbf{E}^2$ *is called an* affine transformation *if there is an invertible* 2 *by* 2 *matrix* A *and a vector* $b \in \mathbf{R}^2$ *such that for all* $x \in \mathbf{R}^2$, $Tx = Ax + b$.

Given a collineation T, we may use Theorem 2.8 to find an affine transformation S that has the same effect on 0, ε_1, and ε_2. Then $S^{-1}T$ is a collineation that leaves 0, ε_1, and ε_2 fixed. If we can prove that $S^{-1}T$ must be the identity, we will be finished.

Before stating the main result, we prove two lemmas.

Lemma 1. *Let* f *be a collineation with* $f(0) = 0$. *If* v *and* w *are nonproportional vectors, then*

$$f(v + w) = f(v) + f(w).$$

Proof: Because $v + w$ is the intersection of $\ell = v + [w]$ and $m = w + [v]$, $f(v + w)$ must be the intersection of $f(\ell)$ and $f(m)$. On the other hand, $f(\ell)$ passes through $f(v)$ and is parallel to $f([w]) = [f(w)]$, whereas $f(m)$ passes through $f(w)$ and is parallel to $[f(v)]$. Because $f(v) + f(w)$ satisfies both of these conditions, it must be the unique point of intersection of $f(\ell)$ and $f(m)$. Thus, $f(v + w)$ and $f(v) + f(w)$ are the same point.

Remark: Lemma 1 is used in the proof of Theorem 1E. Our proof relies on the fact that ℓ and m are not parallel. Once we have proven the main result of this appendix, we will see that Lemma 1 holds true without the restriction on v and w.

Lemma 2. *Let* $\varphi \colon \mathbf{R} \to \mathbf{R}$ *be a bijection satisfying the conditions*

$$\varphi(s + t) = \varphi(s) + \varphi(t), \quad \varphi(st) = \varphi(s)\varphi(t)$$

for all real s, t. *Then* φ *is the identity function.*

Proof: First, note that the given conditions imply that $\varphi(0) = 0$, $\varphi(1) = 1$. Furthermore, $\varphi(-1) = -1$. We can prove easily by induction that $\varphi(n) = n$ for all positive integers n, and, hence, $\varphi(-n) = \varphi((-1)n) = \varphi(-1)\varphi(n) = -n$. Now if $q = m/n$ is a rational number, $\varphi(m) = \varphi(nq) = \varphi(n)\varphi(q)$, so that $\varphi(q) = m/n$. We conclude that φ is the identity on rational numbers.

Next, we show that φ must preserve order in $\mathbf{R}$. First, if a is positive, then $a = b^2$ for some real number b. Then

$$\varphi(a) = \varphi(b^2) = (\varphi(b))^2 > 0.$$

Now, if $t > s$, then $t - s > 0$, so that

$$0 < \varphi(t - s) = \varphi(t) + \varphi(-s) = \varphi(t) - \varphi(s).$$

Suppose now that for some real number t, we have $\varphi(t) > t$. Choose a rational number q between t and $\varphi(t)$ so that

$$t < q < \varphi(t).$$

Because φ preserves order, we must have

$$\varphi(t) < \varphi(q) = q,$$

which is a contradiction. We are forced to conclude that $\varphi(t) > t$ can never occur. The condition $\varphi(t) < t$ would lead to the same conclusion. Hence, $\varphi(t) = t$, and φ is the identity. ☐

Theorem 1E. *Let* f *be a collineation such that* $f(0) = 0$, $f(\varepsilon_1) = \varepsilon_1$, *and* $f(\varepsilon_2) = \varepsilon_2$. *Then* f *is the identity.*

Proof: Because the x_1-axis is a fixed line, there is a function φ such that $f(t\varepsilon_1) = \varphi(t)\varepsilon_1$ for all real t. Similarly, there is a function ψ such that $f(t\varepsilon_2) = \psi(t)\varepsilon_2$. But now, $f(t\varepsilon_1 + t\varepsilon_2) = \varphi(t)\varepsilon_1 + \psi(t)\varepsilon_2$. Because the line $x_2 = x_1$ is fixed, we must have $\varphi = \psi$. But now

$$f(t\varepsilon_1 + s\varepsilon_2) = f(t\varepsilon_1) + f(s\varepsilon_2) = \varphi(t)\varepsilon_1 + \varphi(s)\varepsilon_2. \tag{E.1}$$

In particular, for real m

$$f(\varepsilon_1 + m\varepsilon_2) = \varepsilon_1 + \varphi(m)\varepsilon_2,$$

$$f(x\varepsilon_1 + xm\varepsilon_2) = \varphi(x)\varepsilon_1 + \varphi(xm)\varepsilon_2.$$

Because $(1, m)$, (x, xm), and $(0, 0)$ are collinear, so are $(1, \varphi(m))$, $(\varphi(x), \varphi(xm))$, and $(0, 0)$. Thus, $\varphi(xm) = \varphi(x)\varphi(m)$. In particular, $(\varphi(-1))^2 = \varphi(1) = 1$, so that $\varphi(-1) = -1$.

We will now show that φ also satisfies the identity

$$\varphi(t + s) = \varphi(t) + \varphi(s). \tag{E.2}$$

First, note that

$$f((t + s)\varepsilon_1) = \varphi(t + s)\varepsilon_1.$$

On the other hand,

$$f(t\varepsilon_1 + s\varepsilon_1) = f(t\varepsilon_1 + s\varepsilon_2 + s\varepsilon_1 - s\varepsilon_2)$$
$$= f(t\varepsilon_1 + s\varepsilon_2) + f(s\varepsilon_1 - s\varepsilon_2)$$

by Lemma 1, provided that $t \neq -s$. Again applying Lemma 1, we see that this quantity is equal to

$$f(t\varepsilon_1) + f(s\varepsilon_2) + f(s\varepsilon_1) + f(-s\varepsilon_2)$$
$$= \varphi(t)\varepsilon_1 + \varphi(s)\varepsilon_2 + \varphi(s)\varepsilon_1 + \varphi(-s)\varepsilon_2$$
$$= \varphi(t)\varepsilon_1 + \varphi(s)\varepsilon_1$$

because $\varphi(-s) = \varphi(-1)\varphi(s) = -\varphi(s)$. This completes the proof of (E.2).

An application of Lemma 2 shows that φ is the identity. Hence, by (E.1), f is the identity. $\square$

Corollary (Theorem 2.2). *Every collineation of $\mathbf{E}^2$ is an affine transformation.*

Trigonometric and hyperbolic functions

In this appendix we collect the facts about a special class of functions – the trigonometric and hyperbolic functions – that we use in our geometrical work. A rigorous treatment based on the fundamental properties of the real number system can be found in Kitchen [20] pp. 385–439 or Lang [21], pp. 65–77.

Sine and cosine

The *cosine* function is the unique function satisfying the differential equation

$$f''(x) + f(x) = 0, \quad f(0) = 1, \quad f'(0) = 0.$$

The *sine* function is the unique solution to

$$f''(x) + f(x) = 0, \quad f(0) = 0, \, f'(0) = 1.$$

These functions have domain **R** and range $[-1, 1]$. Their graphs are shown in Figure F.1. (The words "sine" and "cosine" are usually abbreviated sin and cos.) The smallest positive number x such that $\sin x = 0$ is the

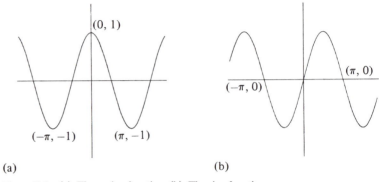

(a)   (b)

Figure F.1   (a) The cosine function (b) The sine function.

familiar number π. In fact, this could be used as the definition of π. The following identities are satisfied by these functions:

$$\cos(\theta + \varphi) = \cos\theta\cos\varphi - \sin\theta\sin\varphi,$$
$$\sin(\theta + \varphi) = \sin\theta\cos\varphi + \cos\theta\sin\varphi. \tag{F.1}$$

$$\cos(-\theta) = \cos\theta, \quad \sin(-\theta) = -\sin\theta, \tag{F.2}$$

$$\cos^2\theta + \sin^2\theta = 1. \tag{F.3}$$

From these, we can easily deduce

$$\cos(\theta - \varphi) = \cos\theta\cos\varphi + \sin\theta\sin\varphi,$$
$$\sin(\theta - \varphi) = \sin\theta\cos\varphi - \cos\theta\sin\varphi, \tag{F.4}$$

$$\cos 2\theta = \cos^2\theta - \sin^2\theta = 2\cos^2\theta - 1 = 1 - 2\sin^2\theta,$$
$$\sin 2\theta = 2\sin\theta\cos\theta. \tag{F.5}$$

The cosine function is decreasing on the interval $[0, \pi]$. It is thus a bijection of $[0, \pi]$ onto $[-1, 1]$.

The inverse bijection is called the *arccos* or $\cos^{-1}$ function. Its graph is shown in Figure F.2. Note that $u = \cos^{-1} v$ if and only if $v = \cos u$ and $0 \leqslant u \leqslant \pi$.

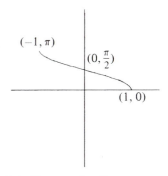

Figure F.2 The arccos function.

Theorem 1F. *Let a and b be numbers such that $a^2 + b^2 = 1$. Then there is a unique number θ in $(-\pi, \pi]$ such that $a = \cos\theta$ and $b = \sin\theta$.*

Proof: Let $u = \cos^{-1} a$. Then $\cos u = a$. Also $\sin u = (1 - \cos^2 u)^{1/2} = (1 - a^2)^{1/2} = |b|$. Note that $u \in [0, \pi]$. If $b > 0$. choose $\theta = u$. Otherwise, choose $\theta = -u$. Then $\cos\theta = a$ and $\sin\theta = b$. For uniqueness, note that any solution θ lies in $(-\pi, 0)$ if $b < 0$ and in $(0, \pi)$ if $b > 0$. Because the cosine function is injective on each of these intervals, only one value of θ is possible. When $b = 0$, θ must be $-\pi$ or 0. One of these is correct for $a = -1$; the other is correct for $a = 1$. □

Theorem 2F. $\cos^{-1} a + \cos^{-1}(-a) = \pi$ *for all* $a \in [-1, 1]$.

Proof: We compute
$$\cos(\pi - \cos^{-1}(-a)) = (\cos\pi)(-a) + (\sin\pi)\sin(\cos^{-1}(-a))$$
$$= (-1)(-a) = a.$$

Because $\pi - \cos^{-1}(-a)$ is a number in $[0, \pi]$ whose cosine is equal to a, we must conclude that it is equal to $\cos^{-1} a$. □

The following lemma is used in Chapter 4 to establish uniqueness of parametrizations of lines of $\mathbf{S}^2$.

Lemma. *Let* $\alpha(t) = (\cos t, \sin t)$. *Suppose that I and I' are closed intervals of length $< 2\pi$ such that $\alpha(I) = \alpha(I')$. Then $I \equiv I' \bmod 2\pi$.*

Proof: Moving each interval by a multiple of 2π if necessary, we may assume that $I = [a, b]$ and $I' = [a', b']$, where $0 \le a, a' \le 2\pi$. Note that $a' \le b$, because, otherwise, $b < a' < 2\pi$ and $\alpha(a')$ cannot be in $\alpha(I)$. By symmetry we also have $a \le b'$.

Assume now that $a < a'$. Choose $u > 0$ so that $u < a' - a$ and $u < 2\pi - (b' - a')$. Then $a < a' - u < b$, so there is a number $v \, \varepsilon \, I'$ such that $\alpha(a' - u) = \alpha(v)$. This implies that $a' - u \equiv v \bmod 2\pi$, which is impossible in light of the inequalities

$$a' - u < a' \le v \quad \text{and} \quad a' - u > b' - 2\pi \ge v - 2\pi.$$

By symmetry, $a' < a$ is also impossible, so $a = a'$. $\qquad\square$

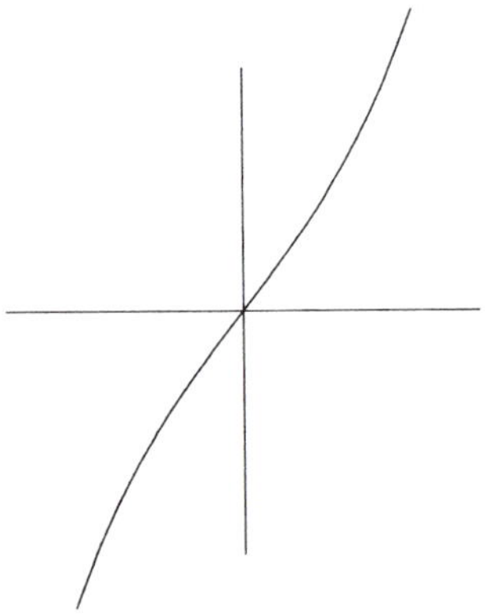

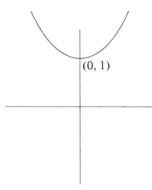

Figure F.3 The hyperbolic sine.

Sinh and cosh

The hyperbolic cosine function *cosh* is the unique function satisfying

$$f''(x) - f(x) = 0, \quad f(0) = 1, \quad f'(0) = 0,$$

and the hyperbolic sine function *sinh* is determined by

$$f''(x) - f(x) = 0, \quad f(0) = 0, \quad f'(0) = 1.$$

The hyperbolic sine is a bijection of **R** onto **R** whose graph is shown in Figure F.3; cosh maps **R** onto $[1, \infty)$, is decreasing on $(-\infty, 0]$, and increasing on $[0, \infty)$. Its shape is that of the familiar "hanging cable." See Figure F.4. The following identities are satisfied:

$$\cosh(u + v) = \cosh u \cosh v + \sinh u \sinh v,$$
$$\sinh(u + v) = \sinh u \cosh v + \cosh u \sinh v, \tag{F.6}$$

$$\cosh(-u) = \cosh u, \quad \sinh(-u) = -\sinh u, \tag{F.7}$$

$$\cosh^2 u - \sinh^2 u = 1. \tag{F.8}$$

Figure F.4 The hyperbolic cosine.

From these follow

$$\cosh(u - v) = \cosh u \cosh v - \sinh u \sinh v,$$
$$\sinh(u - v) = \sinh u \cosh v - \cosh u \sinh v, \tag{F.9}$$

and

$$\cosh 2u = \cosh^2 u + \sinh^2 u$$
$$= 2\cosh^2 u - 1 = 1 + 2\sinh^2 u, \tag{F.10}$$

$$\sinh 2u = 2\sinh u \cosh u.$$

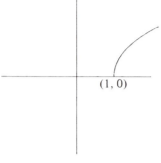

Figure F.5 The arccosh function.

When restricted to the interval $[0, \infty)$, cosh provides a bijection of this interval onto $[1, \infty)$. The inverse bijection is called the *arccosh* or $\cosh^{-1}$ function. See Figure F.5. Note that $u = \cosh^{-1}v$ if and only if $v = \cosh u$ and $u \geq 0$. The domain of $\cosh^{-1}$ is $[1, \infty)$.

Theorem 3F. *Let a and b be numbers such that $a^2 - b^2 = 1$. Then there is a unique number u such that $a = \pm\cosh u$ and $b = \sinh u$.*

Proof: Because sinh is a bijection, the equation $b = \sinh u$ has a unique solution. For this value of u, $\cosh^2 u = 1 + \sinh^2 u = 1 + b^2 = a^2$, so $\cosh u = \pm a$ as required. $\qquad\qquad\qquad\square$

References

1. Abelson, H. and diSessa, A., *Turtle Geometry, The Computer as a Medium for Exploring Mathematics,* Cambridge, MA: M.I.T. Press, 1981.
2. Alperin, J. L., "Groups and Symmetry." In *Mathematics Today*, edited by L. A. Steen. New York: Springer-Verlag, 1978.
3. Banchoff, T. and Wermer, J., *Linear Algebra through Geometry*, New York: Springer-Verlag, 1983.
4. Benson, C. T. and Grove, L. C., *Finite Reflection Groups*, Tarrytown-on-Hudson, NY: Bogdon & Quigley, 1971.
5. Birman, G. and Nomizu, K., Trigonometry in Lorentzian geometry, *Amer. Math. Monthly* 91 (1984), 543–549.
6. Cecil, T. E. and Ryan, P. J., Distance functions and umbilic submanifolds of hyperbolic space, *Nagoya Math. J.* 74 (1979), 67–75.
7. Coxeter, H. S. M., *The Real Projective Plane*, 2d ed., London: Cambridge University Press, 1955.
8. Coxeter, H. S. M., *Introduction to Geometry*, New York: Wiley, 1961.
9. Coxeter, H. S. M., *Non–Euclidean Geometry*, 5th ed., Toronto: University of Toronto Press, 1968.
10. Coxeter, H. S. M., *Projective Geometry*, 2d ed., Toronto: University of Toronto Press, 1974.
11. Eccles, F., *An Introduction to Transformational Geometry*, Reading, MA: Addison-Wesley, 1971.
12. Ewald, G., *Geometry, an Introduction*, Belmont, CA: Wadsworth, 1971.
13. Faber, R. L., *Foundations of Euclidean and non-Euclidean Geometry*, New York: Marcel Dekker, 1983.
14. Foley, J. D. and van Dam, A., *Fundamentals of Interactive Computer Graphics*, Reading, MA: Addison-Wesley, 1982.
15. Frankel, T., *Gravitational Curvature, an Introduction to Einstein's Theory*, San Francisco: Freeman, 1979.
16. Greenberg, M. J., *Euclidean and non-Euclidean Geometries, Development and History*, 2d ed., San Francisco: Freeman, 1980.
17. Guggenheimer, H. W., *Plane Geometry and Its Groups*, San Francisco: Holden-Day, 1967.
18. Heath, T. L., *The Thirteen Books of Euclid's Elements*, 2d ed., 3 vols., New York: Dover 1956.

19. Hoffman, K. and Kunze, R., *Linear Algebra*, 2d ed., Englewood Cliffs, NJ: Prentice-Hall, 1971.

20. Kitchen, J. W. Jr., *Calculus of One Variable*, Reading, MA: Addison-Wesley, 1971.

21. Lang, S., *Analysis* I, Reading, MA: Addison-Wesley, 1968.

22. Martin, G. E., *Transformation Geometry, an Introduction to Symmetry*, New York: Springer-Verlag, 1982.

23. Meschkowski, H. *Noneuclidean Geometry* (translated by A. Shenitzer), New York: Academic Press, 1964.

24. Millman, R. S. and Parker, G. D., *Geometry, a Metric Approach with Models*, New York: Springer-Verlag, 1981.

25. Milnor, J., Hyperbolic geometry: The first 150 years, *Bull. Amer. Math. Soc. (New Series)* 6 (1982), 9–24.

26. Nomizu, K., *Fundamentals of Linear Algebra*, New York: McGraw-Hill, 1966, 2d ed., New York: Chelsea, 1979.

27. Penrose, R., "The Geometry of the Universe." In *Mathematics Today*, edited by L. A. Steen, New York: Springer-Verlag, 1978.

28. Rucker, R. von B., *Geometry, Relativity and the Fourth Dimension*, New York: Dover, 1977.

29. Taylor, E. F. and Wheeler, J. A., *Spacetime Physics*, San Francisco: Freeman, 1966.

30. Tietze, H., *Famous Problems of Mathematics*, New York: Graylock, 1965.

31. Weyl, H., *Symmetry*, Princeton: Princeton University Press, 1952.

32. Thurston, W. P., Three dimensional manifolds, Kleinian groups and hyperbolic geometry, *Bull. Amer. Math. Soc. (New Series)* 6 (1982), 357–381.

33. Veblen, O. and Young, J. W., *Projective Geometry*, 2 vols., Boston: Ginn, 1910, 1918.

34. Yaglom, I. M., *A Simple non-Euclidean Geometry and Its Physical Basis*, New York: Springer-Verlag, 1979.

35. Yale, P. B., *Geometry and Symmetry*, San Francisco: Holden-Day, 1968.

36. Yen, B. L. and Huang, T. S., Determining 3-D motion and structure of a rigid body using the spherical projection, *Computer Vision, Graphics and Image Processing* 21 (1983), 21–32.

Index

Index